Lecture Notes in Computer Science

Commenced Publication in 1973
Founding and Former Series Editors:
Gerhard Goos, Juris Hartmanis, and Jan van Leeuwen

Edoardo Airoldi   David M. Blei
Stephen E. Fienberg   Anna Goldenberg
Eric P. Xing   Alice X. Zheng (Eds.)

Statistical Network Analysis: Models, Issues, and New Directions

ICML 2006 Workshop
on Statistical Network Analysis
Pittsburgh, PA, USA, June 29, 2006
Revised Selected Papers

 Springer

Volume Editors

Edoardo Airoldi
Carnegie Mellon University, Pittsburgh, PA 15213, USA
E-mail: eairlodi@cs.cmu.edu

David M. Blei
Princeton University,Princeton University, USA
E-mail: blei@cs.Princton.edu

Stephen E. Fienberg
Carnegie Mellon University, Pittsburgh, PA 15213-3890, USA
E-mail: fienberg@stat.cmu.edu

Anna Goldenberg
Carnegie Mellon University, Pittsburgh PA 15213-3891, USA
E-mail: anya@cmu.edu

Eric P. Xing
Carnegie Mellon University, Pittsburgh, PA 15217, USA
E-mail: epxing@cs.cmu.edu

Alice X. Zheng
University of California, Berkeley, CA 94720-4, USA
E-mail: alicez@cs.cmu.edu

Library of Congress Control Number: 2007931600

CR Subject Classification (1998): C.2.4, C.2, C.4, E.1, E.3, G.3, I.2.8

LNCS Sublibrary: SL 5 – Computer Communication Networks and
Telecommunications

ISSN   0302-9743
ISBN-10   3-540-73132-6 Springer Berlin Heidelberg New York
ISBN-13   978-3-540-73132-0 Springer Berlin Heidelberg New York

Springer is a part of Springer Science+Business Media

springer.com

© Springer-Verlag Berlin Heidelberg 2007
Printed in Germany

Typesetting: Camera-ready by author, data conversion by Scientific Publishing Services, Chennai, India
Printed on acid-free paper   SPIN: 12078592   06/3180   5 4 3 2 1 0

Preface

This volume was prepared to share with a larger audience the exciting ideas and work presented at an ICML 2006 workshop of the same title.

Network models have a long history. Sociologists and statisticians made major advances in the 1970s and 1980s, culminating in part with a number of substantial databases and the class of exponential random graph models and related methods in the early 1990s. Physicists and computer scientists came to this domain considerably later, but they enriched the array of models and approaches and began to tackle much larger networks and more complex forms of data. Our goal in organizing the workshop was to encourage a dialog among people coming from different disciplinary perspectives and with different methods, models, and tools.

Both the workshop and the editing of the proceedings was a truly collaborative effort on behalf of all six editors, but three in particular deserve special recognition. Anna Goldenberg and Alice Zheng were the driving force behind the entire enterprise and Edo Airoldi assisted on a number of the more important arrangements.

The editing process involved two stages. We were assisted in the review of initial submissions by a Program Committee that including the following individuals:

- David Banks, *Duke University*
- Peter Dodds, *Columbia University*
- Lise Getoor, *University of Maryland*
- Mark Handcock, *University of Washington, Seattle*
- Peter Hoff, *University of Washington, Seattle*
- David Jensen, *University of Massachusetts, Amherst*
- Alan Karr, *National Institute of Statistical Sciences*
- Jon Kleinberg, *Cornell University*
- Andrew McCallum, *University of Massachusetts, Amherst*
- Foster Provost, *New York University*
- Cosma Shalizi, *Carnegie Mellon University*
- Padhraic Smyth *University of California, Irvine*
- Josh Tenenbaum, *Massachusetts Institute of Technology*
- Stanley Wasserman, *Indiana University*

Following the workshop, all papers went through a second round of review and editing (and a few went through a third round).

We are indebted to Caroline Sheedy and Heidi Sestrich, who managed the preparation of the final LaTeX manuscript and did a final cleaning and editing of all material. Without their assistance this volume would not exist.

February 2007   Stephen E. Fienberg

Table of Contents

Part I Invited Presentations

Part II Other Presentations

Part III Extended Abstracts

Part IV Panel Discussion

Structural Inference of Hierarchies in Networks

Aaron Clauset[1], Cristopher Moore[1,2], and Mark E. J. Newman[3]

[1] Department of Computer Science and
[2] Department of Physics and Astronomy,
University of New Mexico, Albuquerque, NM 87131 USA
[3] Department of Physics and Center for the Study of Complex Systems,
University of Michigan, Ann Arbor, MI 48109 USA
aaronc@santafe.edu, moore@cs.unm.edu, mejn@umich.edu

Abstract. One property of networks that has received comparatively little attention is hierarchy, i.e., the property of having vertices that cluster together in groups, which then join to form groups of groups, and so forth, up through all levels of organization in the network. Here, we give a precise definition of hierarchical structure, give a generic model for generating arbitrary hierarchical structure in a random graph, and describe a statistically principled way to learn the set of hierarchical features that most plausibly explain a particular real-world network. By applying this approach to two example networks, we demonstrate its advantages for the interpretation of network data, the annotation of graphs with edge, vertex and community properties, and the generation of generic null models for further hypothesis testing.

1 Introduction

Networks or graphs provide a useful mathematical representation of a broad variety of complex systems, from the World Wide Web and the Internet to social, biochemical, and ecological systems. The last decade has seen a surge of interest across the sciences in the study of networks, including both empirical studies of particular networked systems and the development of new techniques and models for their analysis and interpretation [1, 2].

Within the mathematical sciences, researchers have focused on the statistical characterization of network structure, and, at times, on producing descriptive generative mechanisms of simple structures. This approach, in which scientists have focused on statistical summaries of network structure, such as path lengths [3, 4], degree distributions [5], and correlation coefficients [6], stands in contrast with, for example, the work on networks in the social and biological sciences, where the focus is instead on the properties of individual vertices or groups. More recently, researchers in both areas have become more interested in the global organization of networks [7, 8].

One property of real-world networks that has received comparatively little attention is that of *hierarchy*, i.e., the observation that networks often have a fractal-like structure in which vertices cluster together into groups that then join

E.M. Airoldi et al. (Eds.): ICML 2006 Ws, LNCS 4503, pp. 1–13, 2007.

to form groups of groups, and so forth, from the lowest levels of organization up to the level of the entire network. In this paper, we offer a precise definition of the notion of hierarchy in networks and give a generic model for generating networks with arbitrary hierarchical structure. We then describe an approach for learning such models from real network data, based on maximum likelihood methods and Markov chain Monte Carlo sampling. In addition to inferring global structure from graph data, our method allows the researcher to annotate a graph with community structure, edge strength, and vertex affiliation information.

At its heart, our method works by sampling hierarchical structures with probability proportional to the likelihood with which they produce the input graph. This allows us to contemplate the ensemble of random graphs that are statistically similar to the original graph, and, through it, to measure various average network properties in manner reminiscent of Bayesian model averaging. In particular, we can

1. search for the maximum likelihood hierarchical model of a particular graph, which can then be used as a *null model* for further hypothesis testing,
2. derive a consensus hierarchical structure from the ensemble of sampled models, where hierarchical features are weighted by their likelihood, and
3. annotate an edge, or the absence of an edge, as "surprising" to the extent that it occurs with low probability in the ensemble.

To our knowledge, this method is the only one that offers such information about a network. Moreover, this information can easily be represented in a human-readable format, providing a compact visualization of important organizational features of the network, which will be a useful tool for practitioners in generating new hypotheses about the organization of networks.

2   Hierarchical Structures

The idea of hierarchical structure in networks is not new; sociologists, among others, have considered the idea since the 1970s. For instance, the method known as *hierarchical clustering* groups vertices in networks by aggregating them iteratively in a hierarchical fashion [9]. However, it is not clear that the hierarchical structures produced by these and other popular methods are unbiased, as is also the case for the hierarchical clustering algorithms of machine learning [10]. That is, it is not clear to what degree these structures reflect the true structure of the network, and to what degree they are artifacts of the algorithm itself. This conflation of intrinsic network properties with features of the algorithms used to infer them is unfortunate, and we specifically seek to address this problem here.

A hierarchical network, as considered here, is one that divides naturally into groups *and* these groups themselves divide into subgroups, and so on until we reach the level of individual vertices. Such structure is most often represented as a tree or *dendrogram*, as shown, for example, in Figure 1. We formalize this notion precisely in the following way. Let G be a graph with n vertices. A hierarchical organization of G is a rooted binary tree whose leaves are the graph vertices

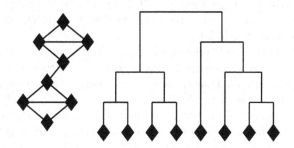

Fig. 1. A small network and one possible hierarchical organization of its nodes, drawn as a dendrogram

and whose internal (i.e., non-leaf) nodes indicate the hierarchical relationships among the leaves. We denote such an organization by $\mathcal{D} = \{D_1, D_2, \ldots, D_{n-1}\}$, where each D_i is an internal node, and every node-pair (u, v) is associated with a unique D_i, their lowest common ancestor in the tree. In this way, $\mathcal{D}$ partitions the edges of G.

3   A Random Graph Model of Hierarchical Organization

We now give a simple model $\mathcal{H}(\mathcal{D}, \boldsymbol{\theta})$ of the hierarchical organization of a network. Our primary assumption is that the edges of G exist independently but with a probability that is not identically distributed. One may think of this model as a variation on the classical Erdős-Rényi random graph, where now the probability that an edge (u, v) exists is given by a parameter θ_i associated with D_i, the lowest common ancestor of u, v in $\mathcal{D}$. Figure 2 shows an example model on seven graph vertices. In this manner, a particular $\mathcal{H}(\mathcal{D}, \boldsymbol{\theta})$ represents an ensemble of inhomogeneous random graphs, where the inhomogeneities are exactly specified by the topological structure of the dendrogram $\mathcal{D}$ and the corresponding Bernoulli trial parameters $\boldsymbol{\theta}$. Certainly, one could write down a more

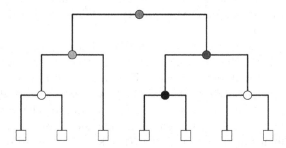

Fig. 2. An example hierarchical model $\mathcal{H}(\mathcal{D}, \boldsymbol{\theta})$, showing a hierarchy among seven graph nodes and the Bernoulli trial parameter θ_i (shown as a gray-scale value) for each group of edges D_i

complicated model of graph hierarchy. The model described here, however, is a relatively generic one that is sufficiently powerful to enrich considerably our ability to learn from graph data.

Now we turn to the question of finding the parametrizations of $\mathcal{H}(\mathcal{D}, \boldsymbol{\theta})$ that most accurately, or rather *most plausibly*, represent the structure that we observe in our real-world graph G. That is, we want to choose $\mathcal{D}$ and $\boldsymbol{\theta}$ such that a graph instance drawn from the ensemble of random graphs represented by $\mathcal{H}(\mathcal{D}, \boldsymbol{\theta})$ will be statistically similar to G. If we already have a dendrogram $\mathcal{D}$, then we may use the method of maximum likelihood [11] to estimate the parameters $\boldsymbol{\theta}$ that achieve this goal. Let E_i be the number of edges in G that have lowest common ancestor i in $\mathcal{D}$, and let L_i (R_i) be the number of leaves in the left- (right-) subtree rooted at i. Then, the maximum likelihood estimator for the corresponding parameter is $\theta_i = E_i/L_i R_i$, the fraction of potential edges between the two subtrees of i that actually appear in our data G. The posterior probability, or likelihood of the model given the data, is then given by

$$\mathcal{L}_{\mathcal{H}}(\mathcal{D}, \boldsymbol{\theta}) = \prod_{i=1}^{n-1} (\theta_i)^{E_i} (1 - \theta_i)^{L_i R_i - E_i} \quad . \tag{1}$$

While it is easy to find values of θ_i by maximum likelihood for each dendrogram, it is not easy to maximize the resulting likelihood function analytically over the space of all dendrograms. Instead, therefore, we employ a Markov chain Monte Carlo (MCMC) method to estimate the posterior distribution by sampling from the set of dendrograms with probability proportional to their likelihood. We note that the number of possible dendrograms with n leaves is super-exponential, growing like $(2n - 3)!! \approx \sqrt{2}\,(2n)^{n-1} e^{-n}$ where !! denotes the double factorial. We find, however, that in practice our MCMC process mixes relatively quickly for networks of up to a few thousand vertices. Finally, to keep our notation concise, we will use $\mathcal{L}_\mu$ to denote the likelihood of a particular dendrogram μ, when calculated as above.

4   Markov Chain Monte Carlo Sampling

Our Monte Carlo method uses the standard Metropolis-Hastings [12] sampling scheme; we now briefly discuss the ergodicity and detailed balance issues for our particular application.

Let ν denote the current state of the Markov chain, which is a dendrogram $\mathcal{D}$. Each internal node i of the dendrogram is associated with three subtrees a, b, and c, where two are its children and one is its sibling—see Figure 3. As the figure shows, these subtrees can be in one of the three hierarchical configurations. To select a candidate state transition $\nu \rightarrow \mu$ for our Markov chain, we first choose an internal node uniformly at random and then choose one of its two alternate configurations uniformly at random. It is then straightforward to show that the ergodicity requirement is satisfied.

Detailed balance is ensured by making the standard Metropolis choice of acceptance probability for our candidate transition: we always accept a transition

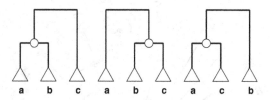

Fig. 3. Each internal dendrogram node i (circle) has three associated subtrees a, b, and c (triangles), which together can be in any of three configurations (up to a permutation of the left-right order of subtrees).

that yields an increase in likelihood or no change, i.e., for which $\mathcal{L}_\mu \geq \mathcal{L}_\nu$; otherwise, we accept a transition that decreases the likelihood with probability equal to the ratio of the respective state likelihoods $\mathcal{L}_\mu / \mathcal{L}_\nu = e^{\log \mathcal{L}_\nu - \log \mathcal{L}_\mu}$. This Markov chain then generates dendrograms μ at equilibrium with probabilities proportional to $\mathcal{L}_\mu$.

5 Mixing Time and Point Estimates

With the formal framework of our method established, we now demonstrate its application to two small, canonical networks: Zachary's karate club [13], a social network of $n = 34$ nodes and $m = 78$ edges representing friendship ties among students at a university karate club; and the year 2000 Schedule of NCAA college (American) football games, where nodes represent college football teams and edges connect teams if they played during the 2000 season, where $n = 115$ and

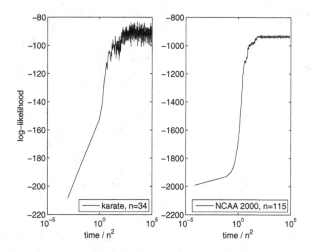

Fig. 4. Log-likelihood as a function of the number of MCMC steps, normalized by n^2, showing rapid convergence to equilibrium

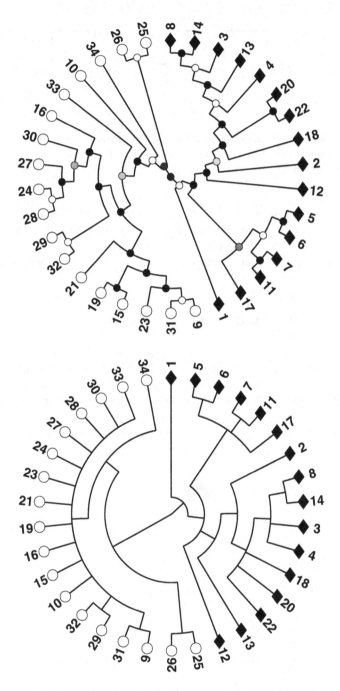

Fig. 5. Zachary's karate club network: (top) an exemplar maximum likelihood dendrogram with $\log \mathcal{L} = -73.32$, parameters θ_i are shown as gray-scale values, and leaf shapes denote group affiliation; and (bottom) the consensus hierarchy sampled at equilibrium. Leaf shapes are common between the two dendrograms, but position varies.

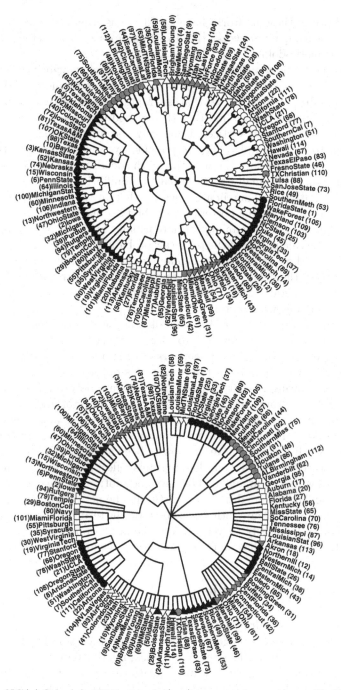

Fig. 6. The NCAA Schedule 2000 network: (top) an exemplar maximum likelihood dendrogram with $\log \mathcal{L} = -884.2$, parameters θ_i are shown as gray-scale values, and leaf shapes denote conference affiliation; and (bottom) the consensus hierarchy sampled at equilibrium. Leaf shapes are common between the two dendrograms, but position varies.

$m = 613$. Both of these networks have found use as standard tests of clustering algorithms for complex networks [14, 15, 16] and serve as a useful comparative basis for our methodology.

Figure 4 shows the convergence of the MCMC sampling algorithm to the equilibrium region of model space for both networks, where we measure the number of steps normalized by n^2. We see that the Markov chain mixes quickly for both networks, and in practice we find that the method works well on networks with up to a few thousands of vertices. Improving the mixing time, so as to apply our method to larger graphs, may be possible by considering state transitions that more dramatically alter the structure of the dendrogram, but we do not consider them here. Additionally, we find that the equilibrium region contains many roughly competitive local maxima, suggesting that any particular maximum likelihood point estimate of the posterior probability is likely to be an overfit of the data. However, formulating an appropriate penalty function for a more Bayesian approach to the calculation of the posterior probability appears tricky given that it is not clear to how characterize such an overfit. Instead, we here compute average features of the dendrogram over the equilibrium distribution of models to infer the most general hierarchical organization of the network. This process is described in the following section.

To give the reader an idea of the kind of dendrograms our method produces, we show instances that correspond to local maxima found during equilibrium sampling for each of our example networks in Figures 5 (top) and 6 (top). For both networks, we can validate the algorithm's output using known metadata for the nodes. During Zachary's study of the karate network, for instance, the club split into two groups, centered on the club's instructor and owner (nodes 1 and 34 respectively), while in the college football schedule teams are divided into "conferences" of 8–12 teams each, with a majority of games being played within conferences. Both networks have previously been shown to exhibit strong community structure [14, 15], and our dendrograms reflect this finding, almost always placing leaves with a common label in the same subtree. In the case of the karate club, in particular, the dendrogram bipartitions the network perfectly according to the known groups. Many other methods for clustering nodes in graphs have difficulty correctly classifying vertices that lie at the boundary of the clusters; in contrast, our method has no trouble correctly placing these peripheral nodes.

6   Consensus Hierarchies

Turning now to the dendrogram sampling itself, we consider three specific structural features, which we average over the set of models explored by the MCMC at equilibrium. First, we consider the hierarchical relationships themselves, adapting for the purpose the technique of *majority consensus*, which is widely used in the reconstruction of phylogenetic trees [17]. Briefly, this method takes a collection of trees $\{T_1, T_2, \ldots, T_k\}$ and derives a majority consensus tree T_{maj} containing only those hierarchical features that have majority weight, where we somehow assign a weight to each tree in the collection. For our purposes, we take the weight of a dendrogram $\mathcal{D}$ simply to be its likelihood $\mathcal{L}_D$, which produces an averaging scheme similar to Bayesian model averaging [10]. Once we have tabulated the majority-weight hierarchical features, we use a reconstruction technique to produce the consensus dendrogram. Note that T_{maj} is always a tree, but is not necessarily strictly binary.

The results of applying this process to our example networks are shown in Figures 5 (bottom) and 6 (bottom). For the karate club network, we observe that the bipartition of the two clusters remains the dominant hierarchical feature after sampling a large number of models at equilibrium, and that much of the particular structure low in the dendrogram shown in Figure 5 (top) is eliminated as distracting. Similarly, we observe some coarsening of the hierarchical structure in the NCAA network, as the relationships between individual teams are removed in favor of conference clusterings.

7   Edge and Node Annotations

We can also assign majority-weight properties to nodes and edges. We first describe the former, where we assign a group affiliation to each node.

Given a vertex, we may ask with what likelihood it is placed in a subtree composed primarily of other members of its group (with group membership determined by metadata as in the examples considered here). In a dendrogram $\mathcal{D}$, we say that a subtree rooted at some node i encompasses a group g if both the majority of the descendants of i are members of group g *and* the majority of members of group g are descendants of i. We then assign every leaf below i the label of g. We note that there may be some leaves that belong to no group,

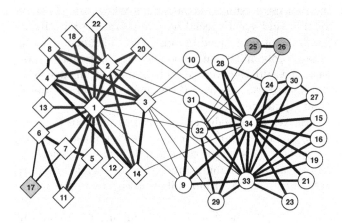

Fig. 7. An annotated version of the karate club network. Line thickness for edges is proportional to their average probability of existing, sampled at equilibrium. Vertices have shapes corresponding to their known group associations, and are shaded according to the sampled weight of their being correctly grouped (see text).

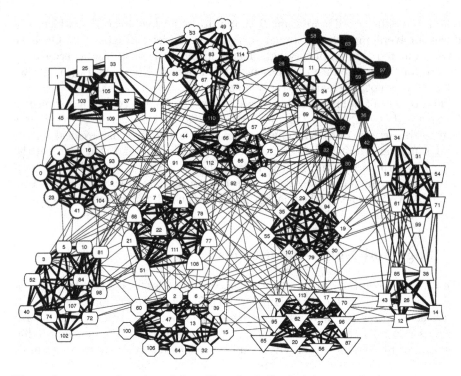

Fig. 8. An annotated version of the college football schedule network. Annotations are as in Figure 7. Note that node shapes here differ from those in Figure 6, but numerical indices remain the same.

i.e., none of their ancestors simultaneously satisfy both the above requirements, and vertices of this kind get a special no-group label. Again, by weighting the group-affiliation vote of each dendrogram by its likelihood, we may measure exactly the average probability that a node belongs to its native group's subtree.

Second, we can measure the average probability that an edge exists, by taking the likelihood-weighted average over the sequence of parameters θ_i associated with that edge at equilibrium.

Estimating these vertex and edge characteristics allows us to annotate the network, highlighting the most plausible features, or the most surprising. Figures 7 and 8 show such annotations for the two example networks, where edge thickness is proportional to average probability, and nodes are shaded proportional to the sampled weight of their native group affiliation (lightest corresponds to highest probability).

For the karate network, the dendrogram sampling both confirms our previous understanding of the network as being composed of two loosely connected groups, and adds additional information. For instance, node 17 and the pair {25, 26} are found to be more loosely bound to their respective groups than other vertices – a feature that is supported by the average hierarchical structure shown in Figure 5 (bottom). This looseness apparently arises because none of these vertices has a direct connection to the central players 1 and 34, and they are thus connected only secondarily to the cores of their clusters. Also, our method correctly places vertex 3 in the cluster surrounding 1, a placement with which many other methods have difficulty.

The NCAA network shows similarly suggestive results, with the majority of heavily weighted edges falling within conferences. Most nodes are strongly placed within their native groups, with a few notable exceptions, such as the independent colleges, vertices 82, 80, 42, 90, and 36, which belong to none of the major conferences. These teams are typically placed by our method in the conference in which they played the most games. Although these annotations illustrate interesting aspects of the NCAA network's structure, we leave a thorough analysis of the data for future work.

8   Discussion and Conclusions

As mentioned in the introduction, we are not the first to study hierarchy in networks. In addition to persistent interest in the sociology community, a number of authors in physics have recently discussed aspects of hierarchical structure [14, 18, 19], although generally via indirect or heuristic means. A closely related, and much studied, concept is that of community structure in networks [14, 15, 16, 20]. In community structure calculations one attempts to find a natural partition of the network that yields densely connected subgraphs or communities. Many algorithms for detecting community structure iteratively divide (or agglomerate) groups of vertices to produce a reasonable partition; the sequence of such divisions (or agglomerations) can then be represented as a dendrogram that is often considered to encode some structure of the graph itself. (Notably, a very recent exception among these community detection heuristics is a method based on maximum likelihood and survey propagation [21].)

Unfortunately, while these algorithms often produce reasonable looking dendrograms, they have the same fundamental problems as traditional hierarchical clustering algorithms for numeric data [10]. That is, it is not clear to what extent the derived hierarchical structures depend on the details of the algorithms used to extract them. It is also unclear how sensitive they are to small perturbations in the graph, such as the addition or removal of a few edges. Further, these algorithms typically produce only a single dendrogram and provide no estimate of the form or number of plausible alternative structures.

In contrast to this previous work, our method directly addresses these problems by explicitly fitting a hierarchical structure to the topology of the graph. We precisely define a general notion of hierarchical structure that is algorithm-independent and we use this definition to develop a random graph model of a hierarchically structured network that we use in a statistical inference context. By sampling via MCMC the set of dendrogram models that are most likely to generate the observed data, we estimate the posterior distribution over models and, through a scheme akin to Bayesian model averaging, infer a set of features that represent the general organization of the network. This approach provides a mathematically principled way to learning about hierarchical organization in real-world graphs. Compared to the previous methods, our approach yields considerable advantages, although at the expense of being more computationally intensive. For smaller graphs, however, for which the calculations described here are tractable, we believe that the insight provided by our methods makes the extra computational effort very worthwhile. In future work, we will explore the extension of our methods to larger networks and characterize the errors the technique can produce.

In closing, we note that the method of dendrogram sampling is quite general and could, in principle, be used to annotate any number of other graph features with information gained by model averaging. We believe that the ability to show which network features are surprising under our model and which are common is genuinely novel and may lead to a better understanding of the inherently stochastic processes that generate much of the network data currently being analyzed by the research community.

Acknowledgments

AC thanks Cosma Shalizi and Terran Lane for many stimulating discussions about statistical inference. MEJN thanks Michael Gastner for work on an early version of the model. This work was funded in part by the National Science Foundation under grants PHY–0200909 (AC and CM) and DMS–0405348 (MEJN) and by a grant from the James S. McDonnell Foundation (MEJN).

References

[1] Albert, R., Barabási, A.L.: Statistical mechanics of complex networks. Rev. Mod. Phys. **74** (2002) 47–97
[2] Newman, M.E.J.: The structure and function of complex networks. SIAM Review **45** (2003) 167–256
[3] Watts, D.J., Strogatz, S.H.: Collective dynamics of 'small-world' networks. Nature **393** (1998) 440–442
[4] Kleinberg, J.: The small-world phenomenon: an algorithmic perspective. In: 32nd ACM Symposium on Theory of Computing. (2000)
[5] Barabási, A.L., Albert, R.: Emergence of scaling in random networks. Science **286** (1999) 509–512
[6] Newman, M.E.J.: Assortative mixing in networks. Phys. Rev. Lett. **89** (2002) 208701
[7] Söderberg, B.: General formalism for inhomogeneous random graphs. Phys. Rev. E **66** (2002) 066121
[8] Wasserman, S., Robins, G.L.: An introduction to random graphs, dependence graphs, and p^*. In Carrington, P., Scott, J., Wasserman, S., eds.: Models and Methods in Social Network Analysis. Cambridge University Press (2005)
[9] Wasserman, S., Faust, K.: Social Network Analysis. Cambridge University Press, Cambridge (1994)
[10] Hastie, T., Tibshirani, R., Friedman, J.: The Elements of Statistical Learning. Springer, New York (2001)

[11] Casella, G., Berger, R.L.: Statistical Inference. Duxbury Press, Belmont (1990)
[12] Newman, M.E.J., Barkema, G.T.: Monte Carlo Methods in Statistical Physics. Clarendon Press, Oxford (1999)
[13] Zachary, W.W.: An information flow model for conflict and fission in small groups. Journal of Anthropological Research **33** (1977) 452–473
[14] Girvan, M., Newman, M.E.J.: Community structure in social and biological networks. Proc. Natl. Acad. Sci. USA **99** (2002) 7821–7826
[15] Radicchi, F., Castellano, C., Cecconi, F., Loreto, V., Parisi, D.: Defining and identifying communities in networks. Proc. Natl. Acad. Sci. USA **101** (2004) 2658–2663
[16] Newman, M.E.J.: Detecting community structure in networks. Eur. Phys. J. B **38** (2004) 321–320
[17] Bryant, D.: A classification of consensus methods for phylogenies. In Janowitz, M., Lapointe, F.J., McMorris, F.R., Mirkin, B., Roberts, F., eds.: BioConsensus. DIMACS (2003) 163–184
[18] Ravasz, E., Somera, A.L., Mongru, D.A., Oltvai, Z.N., Barabási, A.L.: Hierarchical organization of modularity in metabolic networks. Science **30** (2002) 1551–1555
[19] Clauset, A., Newman, M.E.J., Moore, C.: Finding community structure in very large networks. Phys. Rev. E **70** (2004) 066111
[20] Bansal, N., Blum, A., Chawla, S.: Correlation clustering. ACM Machine Learning **56** (2004) 89–113
[21] Hastings, M.B.: Community detection as an inference problem. Preprint cond-mat/0604429 (2006)

Heider vs Simmel:
Emergent Features in Dynamic Structures

David Krackhardt[1] and Mark S. Handcock[2,*]

[1] Carnegie Mellon University, Pittsburgh PA 15213-3890, USA
krack@andrew.cmu.edu
http://www.andrew.cmu.edu/~krack
[2] University of Washington, Seattle WA 98195-4322, USA
handcock@stat.washington.edu
http://www.stat.washington.edu/~handcock

Abstract. Heider's balance theory is ubiquitous in the field of social networks as an explanation for why we so frequently observe symmetry and transitivity in social relations. We propose that Simmelian tie theory could explain the same phenomena without resorting to motivational tautologies that characterize psychological explanations. Further, while both theories predict the same equilibrium state, we argue that they suggest different processes by which this equilibrium is reached. We develop a dynamic exponential random graph model (ERGM) and apply it to the classic panel data collected by Newcomb to empirically explore these two theories. We find strong evidence that Simmelian triads exist and are stable beyond what would be expected through Heiderian tendencies in the data.

1 Heider's Balance Theory

One of the central questions in the field of network analysis is: How do networks form? A cornerstone to our understanding of this process from a structural point of view has been Heider's (1946) theory of balance[1]. According to this theory, a person is motivated to establish and maintain balance in their relationships. What constitutes balance has been the subject of some debate (e.g., [2, 3]), but the core principle has survived and underlies many of our attempts to model this process of network formation (see, for example, [4]).

Heider's (1946) original formulation of balance theory was broad, including people's attitudes towards objects and ideas, not just towards other people. The unifying argument was that people felt comfortable if they agreed with others whom they liked; they felt uncomfortable if they disagreed with others they liked. Moreover, people felt comfortable if they disagreed with others whom they disliked; and people felt uncomfortable if they agreed with others whom they liked.

* We gratefully acknowledge the critical feedback we have received from David Hunter, Steve Goodreau and Carter T. Butts. The research of Handcock was supported by Grant DA012831 from NIDA and Grant HD041877 from NICHD.

E.M. Airoldi et al. (Eds.): ICML 2006 Ws, LNCS 4503, pp. 14–27, 2007.

Heider noted we can represent like and agreement as positive sentiments, and dislike and disagreement as negative sentiments. Considering all combinations of such sentiments among "entities", be they people or objects, Heider simplified the predictions of the theory. "In the case of two entities, a balanced state exists if the relation between them is [mutually] positive (or [mutually] negative.... In the case of three entities, a balanced state exists if all three relations [among the three entities] are positive..., or if two are negative and one positive" (p. 110).

Even in his first paper, Heider noted that in the case where one was considering "entities" as people, then two special properties of balance emerge: symmetry and transitivity. In his terminology, positive affect from one person (p) to another (o) was indicated by "pLo". As noted above, Heider affirms symmetry is basic to balance. His claim for transitivity was more qualified but nonetheless explicit: "Among the many possible cases [of relations among three people, p, o and q] we shall only consider one. $(pLo) + (oLq) + (pLq)$... This example shows ... the psychological transitivity of the L relation [under conditions of balance]" (p. 110).

The other critical tenet in Heider's original formulation was that balance predicted dynamics. Heider's claim was that balance was a state of equilibrium. Imbalance was a state of disequilibrium that would motivate an individual to change something (either a relation or an attitude) that would result in a move toward balance.

It was Cartwright and Harary[5] who first made explicit the connection between Heider's cognitive balance theory and mathematical graph theory. They demonstrated how the principles of balance could be represented by a signed directed graph. Further, by applying the principles of graph theory, they demonstrated how an entire digraph could be characterized as balanced or not depending on the product of the signs of each of its semicycles (or, equivalently, whether semicycles had an even number of negative ties). This extension became the seed for a series of papers and books, each building on Heider's original ideas to study social network structures.

In a series of papers by Leinhardt, Holland and Davis, two critical extensions to this work were developed (see [6], for a spirited review). First, there was the general recognition that most network data, if not actual relations among a set of individuals, were restricted to measurements of positive ties and not negative ties. Thus, they began to look at how balance could be re-thought of as a set of positive-only relations. The concept of transitivity became the dominant theme in these papers. Imbalance was viewed as represented by intransitive triples in the data (cases where $i \rightarrow j$ and $j \rightarrow k$ and not $i \rightarrow k$), rather than the number of negative ties in any semicycle. Balance was viewed as holding if the triple was transitive (or at least vacuously so).

Second, and equally important, they recognized that structures were hardly ever perfectly balanced. The question, they argued, is not whether structures were perfectly balanced but rather whether structures *tended toward* balance, beyond what one would expect by random chance given certain basic features of the graph. They developed a set of distributions and statistical tests for assessing these tendencies and discovered that, indeed, most observed structures show very high degrees of transitivity, relative to chance [7, 8], 1981). This work has remained influential to this day, such that new analyses of balance in any network routinely look at the degree of transitivity (and reciprocity) as measures of balance [4, 9].

2   Simmelian Tie Theory

Simmel, writing at the very start of the 20th century, had a different view of the role of relationships in social settings. He began by noting that the dyad, the fundamental unit of analysis for anyone studying relationships, including social networkers, was *not* the best focus for understanding social behavior. Indeed, he argued that before making any predictions about how two people in a relationship might behave, it is important to understand their context. The context, Simmel continues, is determined by the set of third others who also engage in various relationships with the two focal parties. In other words, Simmel argued that the triad, not the dyad, is the fundamental social unit that needs to be studied.

At the turn of the last century, Simmel provides several theoretical rationales for proffering the triad as the basic social unit ([10]: p. 118-169). Primary among these is that the dyad in isolation has a different character, different set of expectations and demands on its participants, than the dyad embedded in a triad. The presence of a third person changes everything about the dyadic relationship. It is almost irrelevant, according to Simmel, what defines a relationship (marriage, friendship, colleague); Simmel (p. 126-127) even goes so far as to say that "intimacy [the strength or quality of a relationship] is not based on the *content* of the relationship" (emphasis his). Rather, it is based on the structure, the panoply of demands and social dynamics that impinge on that dyad. And those demands are best understood by locating the dyad within its larger context, by finding the groups of people (of at least three persons) that the dyadic members belong to.

Simmel articulates several features that differentiate what he terms the "isolated dyad" from the dyad embedded in a threesome. First, the presence of a third party changes the nature of the relationship itself. Members of a dyad experience an "intensification of relation by [the addition of] a third element, or by a social framework that transcends both members of the dyad" (p. 136).

Similarly, members of a dyad are freer to retain their individuality than members of a group. "[A dyad by itself] favors a relatively greater individuality of its members.... [W]ithin a dyad, there can be no majority which could outvote the individual." (p. 137). Groups, on the other hand, develop norms of behavior; they develop rules of engagement. Individuality is less tolerated in a group, and conformity is more strongly enforced.

Conflict is more easily managed within a triad than in a dyad. Dyadic conflict often escalates out of control. The presence of a third party can ameliorate any conflict, perhaps through mediation, or perhaps simply through diffuse and indirect connection. "Discords between two parties which they themselves cannot remedy are accommodated by the third or by absorption in a comprehensive whole" (p. 135).

Perhaps most central to Simmel's idea about triads is that groups develop an identity, a "super-individual unit" (p. 126). It is a social unit that is larger in meaning and scope than any of its individual components. A consequence of this super-individual identity is that it will outlast its members. That is, people may leave, they may even die, but the group is presumed to carry on. In a triad, the emergent "super-individual unit ... confronts the individual, while at the same time it makes him participate in it" (p. 126). In contrast, dyads by themselves do not reflect this transition to a larger-than-self unit. The dyad's existence is dependent on "the immediacy of interaction" of the two members of the dyad (p. 126). Once one person withdraws from the relationship, the dyad ceases to exist. "A dyad... depends on each of its two elements alone — in its death, though not in its life: for its life, it needs both, but for its death, only one" (p. 124). Thus, he argues, the presence of a third party creates a qualitatively different unit of identity, one that is more stable over time, and one that is more difficult to extricate oneself from.

Finally, Simmel also notes that, while triads are the smallest form of group, increasing group size does not significantly alter its critical features. "[T]he addition of a third person [to dyads] completely changes them, but ... the further expansion to four or more by no means correspondingly modifies the group any further" (p. 138).

Thus, a triad is substantively different from a dyad. The triad is the smallest form of a group. But its existence transforms the nature of all its dyadic constituencies in several important ways. It makes the relationships stronger; it makes them more stable; it makes them more controlling of the behavior of its members.

2.1   Simmelian Ties and Simmelian Decomposition

The foregoing line of Simmelian reasoning suggests that knowing the specific content, nature and strength of a relationship between pairs of people is insufficient to understand the dynamics that might emerge in a social system. Even at the dyadic level, it is critical to know whether any particular dyad is embedded in a group.

To explore the implications of Simmel's theory, Krackhardt[11] proposed using graph theoretic cliques [12] to identify groups. He then defined a Simmelian tie as a tie that was embedded in a clique. Formally, given a directed graph R such that $R_{i,j} = 1$ implies the directed arc $i \rightarrow j$ exists in R, then $R_{i,j}$ is **defined as a Simmelian tie** if and only if the following are all true:

$$R_{i,j} = 1$$

$$R_{j,i} = 1$$

$$\exists k \mid R_{i,k} = 1 \wedge R_{k,i} = 1 \wedge R_{j,k} = 1 \wedge R_{k,j} = 1$$

Gower [13] and more specifically Freeman [14] developed a method of decomposing networks into two components: asymmetric (or specifically "skew-symmetric" in their terminology) and symmetric. Freeman showed that by doing so one could capture more clearly the hierarchy that existed in the network data. Krackhardt extended Freeman's idea by proposing that a directed graph of network ties could be decomposed into three mutually exclusive and exhaustive types: asymmetric, sole-symmetric and Simmelian[11]. These types are defined on a directed graph R:

$$R_{i,j} = \begin{cases} \text{Asymmetric,} & \text{if } R_{i,j} = 1 \land R_{j,i} \neq 1; \\ \text{Sole-Symmetric,} & \text{if } R_{i,j} = 1 \land R_{j,i} = 1 \land R_{i,j} \text{ is not Simmelian;} \\ \text{Simmelian,} & \text{if } R_{i,j} \text{ meets definitional conditions above} \end{cases}$$

2.2   Evidence for the Strength of Simmelian Ties

Since this definition of Simmelian Tie was proposed, several studies have emerged testing various elements of Simmel's theory. Krackhardt [11] re-analyzed the data collected by Newcomb[15] to determine the stability of Simmelian ties relative to asymmetric and sole-symmetric ties. Newcomb had collected network data among a set of 17 college students assigned to live together in a fraternity house. In exchange for reimbursement for living expenses, each student filled out a questionnaire each week for 15 consecutive weeks (except for week 9, where the data were not collected). The network question asked each student to rank order all the remaining 16 students based on how much he liked the others.

For purposes of his analysis, Krackhardt[11] dichotomized these rankings at the median: a relatively high ranking of 1-8 was coded as a 1 (the tie exists); a relatively low ranking of 9-16 was coded as a 0 (the tie does not exist). He then asked the question, which ties have a higher survival rate: asymmetric ties, sole-symmetric ties, or Simmelian ties?

To address this question, he plotted the conditional probabilities that a tie would appear again after Δ weeks, where Δ ranged from 1 week to 14 weeks. That is, given that a tie of a particular type (asymmetric, sole-symmetric, Simmelian) existed at time t, what is the probability that a tie (of any type) will exist at time $t + \Delta$?

His results are reproduced in Figure 1. As can be seen in the graph, ties that were initially embedded in cliques (Simmelian ties) were substantially more likely to survive over time than either asymmetric or sole-symmetric ties. Simmelian ties survived at a rate hovering around .9 for up to 4 weeks, and decay to a rate of near .7 over a 14 week gap. In contrast, both asymmetric ties and sole-symmetric ties survived at a rate of .8 over 1 week's time, dropping quickly to a rate of .7 after 3-4 weeks, and continued down to about .5 after 14 weeks. Clearly, Simmelian contexts provided a substantial survival advantage for ties.

An interesting aside here was that over a large range of time lags (from about 6 to 12 weeks lag), asymmetric ties were considerably more durable than sole-symmetric ties. One possible interpretation of this is that reciprocity in ties, one

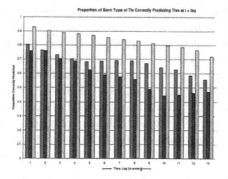

Fig. 1. Stability of Tie Types

of the key elements in Heider's balance theory, led to relative stability only when such ties were embedded in Simmelian triads.

However, Krackhardt did not provide any inferential tests for his results, a shortcoming we will return to later.

A second study[16] explored how much information was contained in Simmelian ties compared to raw ties (un-decomposed ties). The firm being studied

```
          2 3 2 2   2 2 1 3 1 1 3     2   1     2 1 1 2   1 3 1 3 3 1 2 2 1 3
Level     2 7 3 2 2 4 9 0 9 5 3 1 4 4 6 5 1 8 2 1 3 6 0 7 7 4 8 0 9 1 5 3 5 8 6 6
-----     - - - - - - - - - - - - - - - - - - - - - - - - - - - - - - - - - - - -
0.577     . . . . . . . . . . . XXX . . . . . . . . . . . . . . . . . . . . . . . .
0.545     . . . . . . . . . . . XXX . . . . . . . . . . . . . . . . . . . . . . . XXX
0.524     . . XXX . . . . . . . . XXX . . . . . . . . . . . . . . . . . . . . . . . XXX
0.512     . . XXX . . . . . . . . XXX . . . . . . . . . . . . . . . . . . XXX . . XXX
0.473     . . XXX . . . . . . . . XXX . . . . . . . . . XXX . . . . . . . XXX . . XXX
0.473     . . XXX . . . . . XXX . XXX . . . . . . . . . . XXX . . . . . . XXX . . XXX
0.469     . . XXX . . . . . XXX . XXX . . . . . . . XXX XXX . . . . . . . XXX . . XXX
0.450     . . XXX . . . . . XXX XXXXX . . . . . . . XXX XXX . . . . . . . XXX . . XXX
0.442     . . XXX . . . XXX XXX XXXXX . . . . . . . XXX XXX . . . . . . . XXX . . XXX
0.403     . . XXX . . . XXX XXX XXXXX . . . . . . . XXX XXX . . . . . . . XXX XXXXX
0.394     . . XXX . . . XXX XXX XXXXX . . . XXX . XXX XXX . . . . . . . XXX XXXXX
0.393     . . XXX . . . XXX XXX XXXXX . . . XXX . XXX XXX . . . . . . . XXXXXXXXX
0.377     XXX XXX . . . XXX XXX XXXXX . . . XXX . XXX XXX . . . . . . . XXXXXXXXX
0.366     XXX XXX . . . XXX XXX XXXXX . . . XXX . XXX XXXXX . . . . . . XXXXXXXXX
0.366     XXX XXX . . . XXX XXXXXXXXX . . . XXX . XXX XXXXX . . . . . . XXXXXXXXX
0.365     XXX XXX . . . XXX XXXXXXXXX . . . XXX . XXX XXXXX XXX . . . . XXXXXXXXX
0.359     XXX XXX . . . XXX XXXXXXXXX . XXX XXX . XXX XXXXX XXX . . . . XXXXXXXXX
0.358     XXX XXX . . . XXX XXXXXXXXX . XXX XXX . XXX XXXXX XXX . . XXX XXXXXXXXX
0.348     XXX XXX . . . XXXXXXXXXXXXX . XXX XXX . XXX XXXXX XXX . . XXX XXXXXXXXX
0.324     XXXXXXX . . . XXXXXXXXXXXXX . XXX XXX . XXX XXXXX XXX . . XXX XXXXXXXXX
0.323     XXXXXXX . . . XXXXXXXXXXXXX . XXX XXX . XXX XXXXX XXXXX . XXX XXXXXXXXX
0.317     XXXXXXX XXX . XXXXXXXXXXXXX . XXX XXX . XXX XXXXX XXXXX . XXX XXXXXXXXX
0.295     XXXXXXX XXX . XXXXXXXXXXXXX XXXXX XXX . XXX XXXXX XXXXX . XXX XXXXXXXXX
0.283     XXXXXXX XXX . XXXXXXXXXXXXX XXXXX XXX XXXXX XXXXX XXXXX . XXX XXXXXXXXX
0.279     XXXXXXX XXX . XXXXXXXXXXXXX XXXXX XXX XXXXX XXXXX XXXXX . XXXXXXXXXXXXX
0.250     XXXXXXX XXX . XXXXXXXXXXXXX XXXXX XXX XXXXX XXXXX XXXXX XXXXXXXXXXXXX
0.235     XXXXXXX XXX XXXXXXXXXXXXXXX XXXXX XXX XXXXX XXXXX XXXXX XXXXXXXXXXXXXXX
0.224     XXXXXXX XXXXXXXXXXXXXXXXXXX XXXXX XXX XXXXX XXXXX XXXXX XXXXXXXXXXXXXXX
0.216     XXXXXXXXXXXXXXXXXXXXXXXXXXX XXXXX XXXXXXXXX XXXXX XXXXX XXXXXXXXXXXXXXX
0.170     XXXXXXXXXXXXXXXXXXXXXXXXXXX XXXXX XXXXXXXXX XXXXX XXXXX XXXXXXXXXXXXXXX
0.152     XXXXXXXXXXXXXXXXXXXXXXXXXXX XXXXX XXXXXXXXX XXXXXXXXXX XXXXXXXXXXXXXXX
0.140     XXXXXXXXXXXXXXXXXXXXXXXXXXX XXXXX XXXXXXXXXXXXXXX XXXXXXXXXXXXXXXXXXX
0.124     XXXXXXXXXXXXXXXXXXXXXXXXXXX XXXXXXXXXXXXXXXXXXX XXXXXXXXXXXXXXXXXXXXX
0.096     XXXXXXXXXXXXXXXXXXXXXXXXXXX XXXXXXXXXXXXXXXXXXXXXXXXXXXXXXXXXXXXXXXXXXX
```

Fig. 2. Role Analysis Based on Raw Ties

had 36 employees, 15 of which were involved in a unionization effort. Some of the people involved were in favor of the union; some were against it. Some were vocal about their positions; some were quiet. He was able to demonstrate that the dynamics in the union drive and the subsequent defeat of the union was explained by observing how several key supporters of the union were "trapped" in Simmelian ties that kept them from freely expressing their views.

As part of this analysis, Krackhardt examined structurally equivalent role sets for the 36 employees [17]. He noted how clearly these roles emerged in the analysis of the Simmelian relations. What he failed to do was compare these roles to what would have been found had he analyzed the roles uncovered in the raw data.

We have re-analyzed his data (Figures 2 and 3) to make this comparison. Figure 2 provides the dendrogram for the role analysis for the raw data, as is typically done in role analysis in network data. The critical values on the left (vertical axis) represent correlations indicating how similar the roles are that people occupy at that particular cutoff level. What is clear from the analysis of the raw data in this figure is that roles are not coherent. To reach even a modest .3 correlation, the 36 people had to be divided up into 14 different roles. With an average of only a little over 2 people in each role, we learn very little about how role constraints based on the raw data may be playing a part in understanding the union dynamics here.

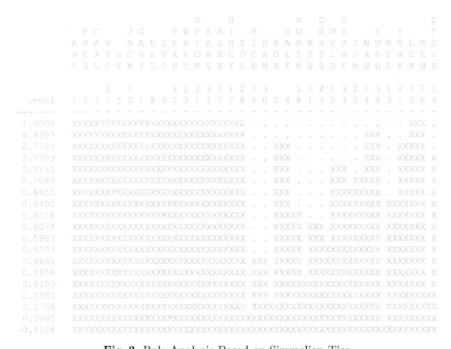

Fig. 3. Role Analysis Based on Simmelian Ties

Figure 3 conducts the same analysis on the Simmelian ties. In contrast to the dendrogram in Figure 2, the correlations show a much better fit with fewer roles. Indeed, collapsing the 36 employees into 5 roles (an average of over 7 people per role) yields an average role similarity correlation of .42, a marked improvement over what was observed in the role analysis for raw data. A reasonable interpretation of these results suggests that raw data are noisy, making them difficult to see systematic patterns of roles and role constraints. Simmelian data appear much cleaner, crisper, suggesting that they could provide the informational backbone for structural analysis.

Thus, we have evidence that Simmelian ties are more stable, and that they provide a stronger, clearer picture of certain structural features in the network. However, again, these are descriptive measures. There is no stochastic model here, and hence no statistical framwork within which we can assess the extent to which these results may be statistical artifacts or perhaps not different from what we would expect by chance. Moreover, these results tell us little about the dynamics of the process of network formation.

3 Dynamic Model Comparison of Heider and Simmel

We return to the central question we started with. What are the forces that seem to help us understand how networks form? We have presented two possible models, competing in their explanations of network dynamics. Both Heider and Simmel are similar in that they "predict" that one should observe many cliques (symmetric and transitive subgraphs). But their motivational underpinnings and their subtle dynamics are radically different.

Heider's model is a psychologically based one. People are motivated to right an imbalance (asymmetric pair or pre-transitive triple) to make it balanced (symmetric and transitive). Once balance is reached, people are said to have reached an equilibrium state and are motivated to maintain that balance. Simmel's theory, by contrast, rests in a sociological, structural explanation for the existence of symmetric and transitive triples. Cliques, once formed, become strong and stable; they resist change. However, there is no inherent motivation to *form* cliques. It's just that, once formed, the ties enter a phase that simultaneously increases their strength and reduces their propensity to decay over time. Thus, one could easily predict an equilibrium for each model that would be the same — dominance of symmetric pairs and transitive triples.

To see which model may better represent the real world, we re-analyzed the Newcomb data. These data provide an opportunity to not only see where the equilibrium might be headed but also to uncover what the actual dynamics are that form the pathway to that equilibrium.

We consider exponential random graph (ERG) models for the network. This class of models allow complex social structure to be represented in an interpretable and parsimonious manner [18, 19]. The model is a statistical exponential family for which the sufficient statistics are a set of functions $Z(r)$ of the network r. The statistics $Z(r)$ are chosen to capture the hypothesized social structure of the network [20]. Models take the form:

$$P_\theta(R = r) = \frac{\exp\left(\theta \cdot Z(r)\right)}{\sum_{s \in \mathcal{R}} \exp\left(\theta \cdot Z(s)\right)}, \tag{1}$$

where $\mathcal{R}$ is the set of all possible networks and θ is our parameter vector. In this form, it is easy to see that $\sum_{s \in \mathcal{R}} \exp\left(\theta \cdot Z(s)\right)$ normalizes our probabilities to ensure a valid distribution. Inference for the model parameter θ can be based on the likelihood function corresponding to the model (1). As the direct computation of the likelihood function is difficult, we approximate it via a MCMC algorithm [21].

The parameter corresponding to a statistic can be interpreted as the log-odds of a tie conditional on the other statistics in the model being fixed. It is also the logarithm of the ratio of the probability of a graph to a graph with a count one lower of the statistic (and all other statistics the same). Hence a positive parameter value indicates that the structural feature occurs with greater frequency than one would expect by chance (all else being fixed). A negative value indicates that the particular structural feature appears less than one would expect by chance.

The space of networks $\mathcal{R}$ we consider for the Newcomb data are those that satisfy the definition of Section 2.2. Each student has exactly 8 out-ties. Hence the density and out-degree distribution of the network are fixed. To capture the propensity for a network to have Heiderian ties and triads we use two statistics:

$$Z_1(r) = \text{number of symmetric dyads in } r \tag{2}$$

Note that the number of edges in the graph is fixed at $17 \times 8 = 136$ and:

$$\text{number of edges} = Z_1 + \text{number of asymmetric dyads} \tag{3}$$

and the total number of dyads is $\binom{17}{2} = 136$ so

$$\text{number of asymmetric dyads} = 136 - Z_1$$
$$\text{number of null dyads} = \frac{1}{2}Z_1$$
$$\text{number of symmetric dyads} = Z_1$$

Hence Z_1 is sufficient to represent the Heiderian dyad census. To represent Heiderian triads we incorporate the statistic:

$$Z_2(r) = \text{number of Heiderian (i.e., transitive) triads in } r$$

To capture the propensity for the network to have Simmelian triads we incorporate the statistic:

$$Z_3(r) = \text{number of Simmelian triads in } r,$$

that is, the number of complete sub-graphs of size three.

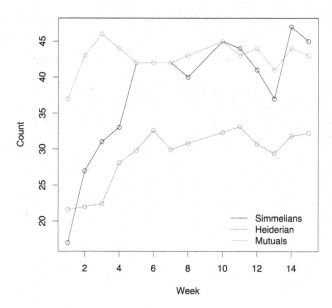

Fig. 4. Simmelian and Heiderian Statistics for the Newcomb Networks over Time. The number of Heiderian triads to divided by two to keep it on a common scale.

A model with this sample space $\mathcal{R}$ controls for density and for individual out-degree patterns.

Figure 4 plots the three statistics for each of the 14 networks. The number of Heiderian triads generally increases over time with a larger rise in the initial weeks. The number of symmetric dyads jumps up, but is not generally increasing or decreasing. The number of Simmelian triads rapidly increases for the first five weeks and then is generally flat pattern. These descriptions can be supported by the confidence intervals for the parameters these statistics represent (not shown).

Traditional ERGM models use such parameters as static structural features. In our case, we are concerned about the transition from a state at time t and the subsequent state at time $t+1$. Thus, we introduce a dynamic variant of the above model:

$$P_\theta(R^{(t+1)} = r^{(t+1)}|R^{(t)} = r^{(t)}) = \frac{\exp\left(\theta^{(t+1)} \cdot Z\left(r^{(t+1)}; r^{(t)}\right)\right)}{\sum_{s\in\mathcal{R}} \exp\left(\theta^{(t+1)} \cdot Z\left(s; r^{(t)}\right)\right)} \qquad t = 2, \ldots, 15,$$

$$(4)$$

where $\mathcal{R}$ is still the set of all possible networks with each student having out-degree four and $\theta^{(t+1)}$ is our parameter vector for the t to $t+1$ transition. The network statistics $Z\left(r^{(t+1)}; r^{(t)}\right)$ indicate how the network (statistics) at time $t+1$ depend on the state at time t. This general model is adapted to the Newcomb data via two additional statistics dynamic statistics:

$Z_4(r^{(t+1)}; r^{(t)})$ = number of pre-Heiderian triads in $r^{(t)}$ that are Heiderian
         in $r^{(t+1)}$

$Z_5(r^{(t+1)}; r^{(t)})$ = number of Simmelian triads in $r^{(t)}$ that persist in $r^{(t+1)}$

The first follows the dynamics of pre-Heiderian (imbalanced) triads from time t to time $t+1$. If there is a Heiderian process evolving we expect to see an increased propensity for the formation of Heiderian triads from their pre-Heiderian states (all else being equal). The second follows the dynamics of Simmelian (complete triples) triads from time t to time $t + 1$. If there is a Simmelian process evolving we expect to see persistence of Simmelian triads (all else being equal). Note that this allows a distinct process of Simmelian formation not controlled by this parameter. By including these statistics in the model, we can follow the dynamics in the Newcomb data to see how states transitioned from a non-balanced state and the stability of Simmelian state once formed.

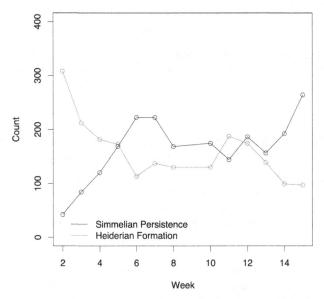

Fig. 5. The persistence of Simmelian triads and the formation of Heiderian triads for the Newcomb Networks over Time

Figure 5 plots the two dynamic statistics over the 14 weeks of data. We clearly see the increase persistence of Simmelian triads over time and the decreasing formation of Heiderian triads over time. Both these effects are strongest in the early weeks with a possible increase in the final weeks.

Both the cross-sectional and dynamic models and figures present overall Heiderian and Simmelian effects. To understand the interactions we consider the joint effects through the parameters of a dynamic model. Consider the model (4)

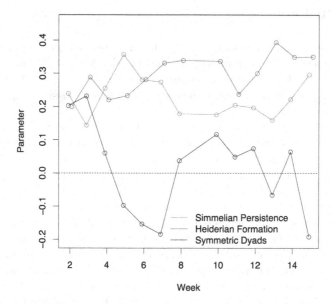

Fig. 6. The joint effects of the persistence of Simmelian triads, Heiderian dyadic balance, and the formation of Heiderian triads for the Newcomb networks over time. The values plotted are the maximum likelihood estimates of the parameters of model (4) for $t = 2, \dots, 15$.

with statistics Z_1, Z_4 and Z_5. These measure the overall level of Heiderian dyads and the dynamics of the two processes. The maximum likelihood estimator of the parameter $\theta^{(t+1)}$ was estimated for $t = 2, \dots, 15$.

Figure 6 plots the parameters over the 14 weeks of data. It is important to note that these measure the simultaneous joint effect of the three factors. Consider the formation of Heiderian triads. We see that is positive for each time point indicating that Heiderian formation is substantively higher than due to chance. It is also modestly increasing over time indicating that the propensity for formation is modestly increasing even in the presence of the other structural factors. The pattern for the Simmelian persistence is also positive indicating substantially more persistence of Simmelian triads than expected due to chance *even adjusting for the Heiderian triadic and dyadic effects.* This has an early peak in the fifth week and appear to be increasing in the last weeks. Both these effects are confirmed by the confidence intervals for the parameters (not shown). Finally, the overall presence of Heiderian dyads is not significantly different from the random process. This is confirmed by the confidence intervals for the symmetric parameter, and also indicated by the point estimates arranged about zero.

All analyses in this section were implemented using the **statnet** package for network analysis [22]. This is written in the **R** language [23] due to its flexibility and power. **statnet** provides access to network analysis tools for the fitting, plotting, summarization, goodness-of-fit, and simulation of networks. Both **R** and the **statnet** package are publicly available (See websites in the references for details).

4   Conclusion

There have been reams of evidence for the frequent occurrence of symmetric and transitive structures in naturally occurring networks. Most of this work has been motivated by Heider's theory of balance. While Simmel's work is well-known among sociologists, little attention has been paid to his possible explanation of the same phenomena.

We have outlined how Simmel's theory, without resorting to any psychological motivations, can be used to predict the same structures as Heider's theory. Indeed, one would expect the same states from each model in equilibrium. But, the dynamics which reach these final states are substantially different. Statistical evidence from the Newcomb data suggest that Simmel's description of the evolution of these structures is a better fit with the data than Heider's.

The results of the dynamic modeling of the Newcomb data (Figure 6) indicate that Simmelian structures are important to the dynamics in the Newcomb data even when Heiderian dynamics and propensity have been accounted for. Thus the tendency to form Simmelian ties that persist most strongly and significantly throughout time is not just a by-product of a Heiderian process, but exists above and beyond that. The results also indicate that the overall level of Heiderian balance is a product of the dynamic formation of Heiderian triads from pre-Heiderian triads (above and beyond that naturally induced by the numbers of pre-Heiderian triads that exist at that point in time).

The results here are not conclusive. The Newcomb data are limited in their generalizability. But they are suggestive. Perhaps the dynamics that we have attributed all these years to Heider and balance theory are at least in part due to a completely different theory, a structural theory more consistent with Simmel's interpretation of structural dynamics.

References

[1] Heider, F.: Attitudes and cognitive organization. Journal of Psychology **21** (1946) 107–112

[2] Davis, J.A.: Clustering and structural balance in graphs. Human Relations **30** (1967) 181–187

[3] Flament, C.: Independent generalizations of balance. In Holland, P.W., Leinhardt, S., eds.: Perspectives on Social Network Research. Academic Press, New York (1979) 187–200

[4] Doreian, P., Krackhardt, D.: Pre-transitive balance mechanisms for signed networks. Journal of Mathematical Sociology **25** (2001) 43–67

[5] Cartwright, D., Harary, F.: Structural balance: a generalization of heider's theory. Psychological Review **63** (1956) 277–293

[6] Davis, J.A.: The davis /holland /leinhardt studies: An overview. In Holland, P.W., Leinhardt, S., eds.: Perspectives on Social Network Research. Academic Press, New York (1979) 51–62

[7] Holland, P.W., Leinhardt, S.: Some evidence on the transitivity of positive interpersonal sentiment. American Journal of Sociology **72** (1972) 1205–1209

[8] Holland, P.W., Leinhardt, S.: Local structure in social networks. In Heise, D.R., ed.: Sociological Methodology. Jossey-Bass, San Francisco, CA (1975)
[9] Krackhardt, D., Kilduff, M.: Whether close or far: Perceptions of balance in friendship networks in organizations. Journal of Personality and Social Psychology **76** (1999) 770–782
[10] Simmel, G.: The Sociology of Georg Simmel. The Free Press, New York (1908/1950)
[11] Krackhardt, D.: Simmelian ties: Super strong and sticky. In Kramer, R., Neale, M., eds.: Power and Influence in Organizations. Sage, Thousand Oaks, CA (1998) 21–38
[12] Luce, R.D., Perry, A.D.: A method of matrix analysis of group structure. Psychometrika **14** (1949) 95–116
[13] Gower, J.C.: The analysis of asymmetry and orthogonality. In Barra, J., Boudreau, F., Romier, G., eds.: Recent Developments in Statistics. North-Holland Publishing Co., New York (1977) 109–123
[14] Freeman, L.C.: Uncovering organizational hierarchies. Computational and Mathematical Organizational Theory **3**(1) (1997) 5–18
[15] Newcomb, T.N.: The Acquaintance Process. Holt, Rinehart and Winston, New York (1961)
[16] Krackhardt, D.: The ties that torture: Simmelian tie analysis in organizations. Research in the Sociology of Organizations **16** (1999) 183–210
[17] Lorrain, F., White, H.: Structural equivalence of individuals in social networks blockstructures with covariates. Journal of Mathematical Sociology **1** (1971) 49–80
[18] Holland, P.W., Leinhardt, S.: An exponential family of probability distributions for directed graphs. Journal of the American Statistical Association **76**(373) (1981)
[19] Frank, O., Strauss, D.: Markov graphs. Journal of the American Statistical Association **81**(395) (1986) 832–842
[20] Morris, M.: Local rules and global properties: Modeling the emergence of network structure. In Breiger, R., Carley, K., Pattison, P., eds.: Dynamic Social Network Modeling and Analysis. Committee on Human Factors, Board on Behavioral, Cognitive, and Sensory Sciences. National Academy Press: Washington, DC (2003) 174–186
[21] Hunter, D.R., Handcock, M.S.: Inference in curved exponential family models for networks. Journal of Computational and Graphical Statistics **15** (2005) 482–500
[22] Handcock, M.S., Hunter, D.R., Butts, C.T., Goodreau, S.M., Morris, M.: statnet: An R package for the Statistical Modeling of Social Networks. http://csde.washington.edu/statnet. (2003)
[23] R Development Core Team: R: A Language and Environment for Statistical Computing. R Foundation for Statistical Computing, Vienna, Austria. (2006) ISBN 3-900051-07-0.

Joint Group and Topic Discovery
from Relations and Text

Andrew McCallum, Xuerui Wang, and Natasha Mohanty

University of Massachusetts, Amherst, MA 01003, USA
{mccallum,xuerui,nmohanty}@cs.umass.edu

Abstract. We present a probabilistic generative model of entity relationships and textual attributes; the model simultaneously discovers groups among the entities and topics among the corresponding text. Block models of relationship data have been studied in social network analysis for some time, however here we cluster in multiple modalities at once. Significantly, joint inference allows the discovery of groups to be guided by the emerging topics, and vice-versa. We present experimental results on two large data sets: sixteen years of bills put before the U.S. Senate, comprising their corresponding text and voting records, and 43 years of similar data from the United Nations. We show that in comparison with traditional, separate latent-variable models for words or block structures for votes, our Group-Topic model's joint inference improves both the groups and topics discovered. Additionally, we present a non-Markov continouous-time group model to capture shifting group structure over time.

1 Introduction

Research in the field of social network analysis (SNA) has led to the development of mathematical models that discover patterns in interaction between entities [1]. One of the objectives of SNA is to detect salient groups of entities. Group discovery has many applications, such as understanding the social structure of organizations [2] or native tribes [3], uncovering criminal organizations [4], and modeling large-scale social networks in Internet services such as Friendster.com or LinkedIn.com.

Social scientists have conducted extensive research on group detection, especially in fields such as anthropology [3] and political science [5,6]. Recently, statisticians and computer scientists have begun to develop models that specifically discover group memberships [7,8,9,10]. One such model is the stochastic block structures model [9], which discovers the latent structure, groups or classes based on pair-wise relation data. A particular relation holds between a pair of entities (people, countries, organizations, etc.) with some probability that depends only on the class (group) assignments of the entities. The relations between all the entities can be represented with a directed or undirected graph. The class assignments can be inferred from a graph of observed relations or link data using Gibbs sampling [9]. This model is extended in [10] to automatically select an arbitrary number of groups by using a Chinese Restaurant Process prior.

E.M. Airoldi et al. (Eds.): ICML 2006 Ws, LNCS 4503, pp. 28–44, 2007.

The aforementioned models discover latent groups only by examining whether one or more relations exist between a pair of entities. The Group-Topic (GT) model presented in this paper, on the other hand, considers not only the relations between objects but also the attributes of the relations (for example, the text associated with the relations) when assigning group membership.

The GT model can be viewed as an extension of the stochastic block structures model [9,10] with the key addition that group membership is conditioned on a latent variable associated with the attributes of the relation. In our experiments, the attributes of relations are words, and the latent variable represents the topic responsible for generating those words. Unlike previous methods, our model captures the *(language) attributes* associated with interactions between entities, and uses distinctions based on these attributes to better assign group memberships.

Consider a legislative body and imagine its members forging alliances (forming groups), and voting accordingly. However, different alliances arise depending on the topic of the resolution up for a vote. For example, one grouping of the legislators may arise on the issue of taxation, while a quite different grouping may occur for votes on foreign trade. Similar patterns of topic-based affiliations would arise in other types of entities as well, e.g., research paper co-authorship relations between people and citation relations between papers, with words as attributes on these relations.

In the GT model, the discovery of groups is guided by the emerging topics, and the discovery of topics is guided by emerging groups. Both modalities are driven by the common goal of increasing data likelihood. Consider the voting example again; resolutions that would have been assigned the same topic in a model using words alone may be assigned to different topics if they exhibit distinct voting patterns. Distinct word-based topics may be merged if the entities vote very similarly on them. Likewise, multiple different divisions of entities into groups are made possible by conditioning them on the topics.

The importance of modeling the *language* associated with interactions between people has recently been demonstrated in the Author-Recipient-Topic (ART) model [11]. In ART the words in a message between people in a network are generated conditioned on the author, recipients and a set of topics that describes the message. The model thus captures both the network structure within which the people interact as well as the language associated with the interactions. In experiments with Enron and academic email, the ART model is able to discover role similarity of people better than SNA models that consider network connectivity alone. However, the ART model does not explicitly capture groups formed by entities in the network.

The GT model simultaneously clusters entities to groups and clusters words into topics, unlike models that generate topics solely based on word distributions such as Latent Dirichlet Allocation [12]. In this way the GT model discovers salient topics relevant to relationships between entities in the social network— topics which the models that only examine words are unable to detect. Erosheva et al. [13] provide a general formulation for mixed membership, of which LDA is a special case, and they apply it to soft clustering of papers by topics using words from the text and references. In work parallel to ours but different from GT, Airoldi et al. [14] extend the general mixed membership model to also incorporate stochastic blockmodels of the form arising in the network literature. Their application is to protein-protein interactions.

We demonstrate the capabilities of the GT model by applying it to two large sets of voting data: one from US Senate and the other from the General Assembly of the UN. The model clusters voting entities into coalitions and simultaneously discovers topics for word attributes describing the relations (bills or resolutions) between entities. We find that the groups obtained from the GT model are significantly more cohesive (p-value $< .01$) than those obtained from the block structures model. The GT model also discovers new and more salient topics in both the Senate and UN datasets—in comparison with topics discovered by only examining the words of the resolutions, the GT topics are either split or joined together as influenced by the voters' patterns of behavior.

2   Group-Topic Model

The Group-Topic Model is a directed graphical model that clusters entities with relations between them, as well as attributes of those relations. The relations may be either directed or undirected and have multiple attributes. In this paper, we focus on undirected relations and have words as the attributes on relations.

In the generative process for each event (an interaction between entities), the model first picks the topic t of the event and then generates all the words describing the event where each word is generated independently according to a multinomial (discrete) distribution ϕ_t, specific to the topic t. To generate the relational structure of the network, first the group assignment, g_{st} for each entity s is chosen conditionally from a particular multinomial (discrete) distribution θ_t

Table 1. Notation used in this paper

SYMBOL DESCRIPTION

g_{it}	entity i's group assignment in topic t
t_b	topic of an event b
$w_k^{(b)}$	the kth token in the event b
$V_{ij}^{(b)}$	entity i and j's groups behaved same (1) or differently (2) on the event b
S	number of entities
T	number of topics
G	number of groups
B	number of events
V	number of unique words
N_b	number of word tokens in the event b
S_b	number of entities who participated in the event b

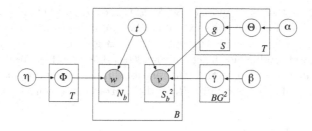

Fig. 1. The Group-Topic model

over groups for each topic t. Given the group assignments on an event b, the matrix $V^{(b)}$ is generated where each cell $V_{ij}^{(b)}$ represents if the groups of two entities (i and j) behaved the same or not during the event b, (e.g., voted the same or not on a bill). Each element of V is sampled from a binomial (Bernoulli) distribution $\gamma_{g_i g_j}^{(b)}$. Our notation is summarized in Table 1, and the graphical model representation of the model is shown in Figure 1.

Without considering the topic of an event, or by treating all events in a corpus as reflecting a single topic, the simplified model (only the right part of Figure 1) becomes equivalent to the stochastic block structures model [9]. To match the block structures model, each event defines a relationship, e.g., whether in the event two entities' groups behave the same or not. On the other hand, in our model a relation may have multiple attributes (which in our experiments are the words describing the event, generated by a per-topic multinomial (discrete) distribution).

When we consider the complete model, the dataset is dynamically divided into T sub-blocks each of which corresponds to a topic. The complete GT model is as follows,

$$t_b \sim \text{Uniform}(\frac{1}{T})$$
$$w_{it}|\phi_t \sim \text{Multinomial}(\phi_t)$$
$$\phi_t|\eta \sim \text{Dirichlet}(\eta)$$
$$g_{it}|\theta_t \sim \text{Multinomial}(\theta_t)$$
$$\theta_t|\alpha \sim \text{Dirichlet}(\alpha)$$
$$V_{ij}^{(b)}|\gamma_{g_i g_j}^{(b)} \sim \text{Binomial}(\gamma_{g_i g_j}^{(b)})$$
$$\gamma_{gh}^{(b)}|\beta \sim \text{Beta}(\beta).$$

We want to perform joint inference on (text) attributes and relations to obtain topic-wise group memberships. Since inference can not be done exactly on such complicated probabilistic graphical models, we employ Gibbs sampling to conduct inference. Note that we adopt conjugate priors in our setting, and thus we can easily integrate out θ, ϕ and γ to decrease the uncertainty associated with them. This simplifies the sampling since we do not need to sample θ, ϕ and γ at all, unlike in [9]. In our case we need to compute the conditional distribution

$P(g_{st}|\mathbf{w}, \mathbf{V}, \mathbf{g}_{-st}, \mathbf{t}, \alpha, \beta, \eta)$ and $P(t_b|\mathbf{w}, \mathbf{V}, \mathbf{g}, \mathbf{t}_{-b}, \alpha, \beta, \eta)$, where $\mathbf{g}_{-st}$ denotes the group assignments for all entities except entity s in topic t, and $\mathbf{t}_{-b}$ represents the topic assignments for all events except event b. Beginning with the joint probability of a dataset, and using the chain rule, we can obtain the conditional probabilities conveniently. In our setting, the relationship we are investigating is always symmetric, so we do not distinguish R_{ij} and R_{ji} in our derivations (only $R_{ij} (i \leq j)$ remain). Thus

$$P(g_{st}|\mathbf{V}, \mathbf{g}_{-st}, \mathbf{w}, \mathbf{t}, \alpha, \beta, \eta)$$

$$\propto \frac{\alpha_{g_{st}} + n_{tg_{st}} - 1}{\sum_{g=1}^{G}(\alpha_g + n_{tg}) - 1} \prod_{b=1}^{B} \left(I(t_b = t) \prod_{h=1}^{G} \frac{\prod_{k=1}^{2} \prod_{x=1}^{d_{g_{st}hk}^{(b)}} \left(\beta_k + m_{g_{st}hk}^{(b)} - x \right)}{\prod_{x=1}^{\sum_{k=1}^{2} d_{g_{st}hk}^{(b)}} \left((\sum_{k=1}^{2}(\beta_k + m_{g_{st}hk}^{(b)})) - x \right)} \right),$$

where n_{tg} represents how many entities are assigned into group g in topic t, c_{tv} represents how many tokens of word v are assigned to topic t, $m_{ghk}^{(b)}$ represents how many times group g and h vote same ($k = 1$) and differently ($k = 2$) on event b, $I(t_b = t)$ is an indicator function, and $d_{g_{st}hk}^{(b)}$ is the increase in $m_{g_{st}hk}^{(b)}$ if entity s were assigned to group g_{st} than without considering s at all (if $I(t_b = t) = 0$, we ignore the increase in event b). Furthermore,

$$P(t_b|\mathbf{V}, \mathbf{g}, \mathbf{w}, \mathbf{t}_{-b}, \alpha, \beta, \eta)$$

$$\propto \left(\frac{\prod_{v=1}^{V} \prod_{x=1}^{e_v^{(b)}} (\eta_v + c_{t_b v} - x)}{\prod_{x=1}^{\sum_{v=1}^{V} e_v^{(b)}} \left(\sum_{v=1}^{V} (\eta_v + c_{t_b v}) - x \right)} \right)^{\lambda} \prod_{g=1}^{G} \prod_{h=g}^{G} \frac{\prod_{k=1}^{2} \Gamma(\beta_k + m_{ghk}^{(b)})}{\Gamma(\sum_{k=1}^{2}(\beta_k + m_{ghk}^{(b)}))},$$

where $e_v^{(b)}$ is the number of tokens of word v in event b. Note that $m_{ghk}^{(b)}$ is not a constant and changes with the assignment of t_b since it influences the group assignments of all entities that vote on event b. We use a weighting parameter λ to rescale the likelihoods from different modalities, as is also common in speech recognition when the acoustic and language models are combined. The GT model uses information from two different modalities. In general, the likelihood of the two modalities is not directly comparable, since the number of occurrences of each type may vary greatly (e.g., there may be far more pairs of voting entities than word occurrences).

3   Related Work

There has been a surge of interest in models that describe relational data, or relations between entities viewed as links in a network, including recent work in group discovery. One such algorithm, presented by Bhattacharya and Getoor [8], is a bottom-up agglomerative clustering algorithm that partitions links in a network into clusters by considering the change in likelihood that would occur if two clusters were merged. Once the links have been grouped, the entities connected by the links are assigned to groups.

Another model due to Kubica et al. [7] considers both link evidence and attributes on entities to discover groups. The Group Detection Algorithm (GDA) uses a Bayesian network to group entities from two datasets, demographic data describing the entities and link data. Unlike our model, neither of these models [8,7] consider attributes associated with the links between the entities. The model presented in [7] considers attributes of an entity rather than attributes of relations between entities.

The central theme of GT is that it simultaneously clusters entities and attributes on relations (words). There has been prior work in clustering different entities simultaneously, such as information theoretic co-clustering [15], and multi-way distributional clustering using pair-wise interactions [16]. However, these models do not also cluster attributes based on interactions between entities in a network.

In our model, group membership defines pair-wise relations between nodes. The GT model is an enhancement of the stochastic block structures model [9] and the extended model of Kemp et al. [10] as it takes advantage of information from different modalities by conditioning group membership on topics. In this sense, the GT model draws inspiration from the Role-Author-Recipient-Topic (RART) model [11]. As an extension of ART model, RART clusters together entities with similar roles. In contrast, the GT model presented here clusters entities into groups based on their relations to other entities.

Exploring the notion that the behavior of an entity can be explained by its (hidden) group membership, Jakulin and Buntine [17] develop a discrete PCA model for discovering groups in the 108 US Senate. A similar model is developed in [18] that examines group cohesion and voting similarity in the Finnish Parliament. We apply our GT model also to voting data. However, unlike [17,18], since our goal is to cluster entities based on the similarity of their voting patterns, we are only interested in whether a pair of entities voted the same or differently, not their actual yes/no votes. Two resolutions on the same topic may differ only in their goal (e.g., increasing vs. decreasing budget), thus the actual votes on one could be the converse of votes on the other. However, pairs of entities who vote the same on one resolution would tend to vote same on the other resolution. To capture this, we model relations as *agreement* between entities, not the yes/no vote itself. This kind of "content-ignorant" feature is similarly found in some work on web log clustering [19].

There has been a considerable amount of previous work in understanding voting patterns [20,5,6], including research on voting cohesion of countries in the EU parliament [5] and partisanship in roll call voting [6]. In these models roll call

Table 2. Average AI for different models for both Senate and UN datasets. The group cohesion in (joint) GT is significantly better than in (serial) baseline, as well as the block structures model that does not use text at all.

Datasets	Avg. AI for GT	Avg. AI for Baseline	p-value	Block Structures
Senate	0.8294	0.8198	$< .01$	0.7850
UN	0.8664	0.8548	$< .01$	0.7934

data are used to estimate *ideal points* of a legislator (which refers to a legislator's preferred policy in the Euclidean space of possible policies). The models assume that each vote in the roll call data is independent of the remaining votes, i.e., each individual is not connected to anyone else who is voting. However, in reality, legislation is shaped by the coalitions formed by like-minded legislators. The GT model attempts to capture this interaction.

4   Experimental Results

We present experiments applying the GT model to the voting records of members of two legislative bodies: the US Senate and the UN General Assembly. We set $\alpha = 1$, $\beta = 5$, and $\eta = 1$ in all experiments. To make sure of convergence, we run the Markov chains for 10,000 iterations, (which by inspection are stable after a few hundred iterations), and use the topic and group assignments in the last Gibbs sample.

For comparison, we present the results of a baseline method that first uses a mixture of unigrams to discover topics and associate a topic with each resolution, and then runs the block structures model [9] separately on the resolutions assigned to each topic. This baseline approach is similar to the GT model in that it discovers both groups and topics, and has different group assignments on different topics. However, whereas the GT model performs joint inference simultaneously, the baseline performs inference serially. Note that our baseline is still more powerful than the block structures models, since it models the topic associated with each event, and allows the creation of distinct groupings dependent on different topics.

In this paper, we are interested in the quality of both the groups and the topics. In the political science literature, group cohesion is quantified by the *Agreement Index (AI)* [17,18], which measures the similarity of votes cast by members of a group during a particular roll call. The AI for a particular group on a given roll call i is based on the number of group members that vote Yes(y_i), No(n_i) or Abstain(a_i) in the roll call i. Higher AI index means better cohesion.

$$AI_i = \frac{\max\{y_i, n_i, a_i\} - \frac{y_i + n_i + a_i - \max\{y_i, n_i, a_i\}}{2}}{y_i + n_i + a_i}$$

The block structures model assumes that members of a legislative body have the same group affiliations irrespective of the topic of the resolution on vote. However, it is likely that members form their groups based on the topic of the resolution being voted on. We quantify the extent to which a member s switches groups with a *Group Switch Index* (GSI).

$$GSI_s = \sum_{i,j}^{T} \frac{\text{abs}(s_i - s_j)}{|G(s,i)| - 1 + |G(s,j)| - 1}$$

where s_i and s_j are bit vectors of the length of the size of the legislative body. The k_{th} bit of s_i is set if k is in the same group as s on topic i and similarly

Table 3. Top words for topics generated with the mixture of unigrams model on the Senate dataset. The headers are our own summary of the topics.

Economic	Education	Military Misc.	Energy
federal	education	government	energy
labor	school	military	power
insurance	aid	foreign	water
aid	children	tax	nuclear
tax	drug	congress	gas
business	students	aid	petrol
employee	elementary	law	research
care	prevention	policy	pollution

Table 4. Top words for topics generated with the GT model on the Senate dataset. The topics are influenced by both the words and votes on the bills.

Economic	Education + Domestic	Foreign	Social Security + Medicare
labor	education	foreign	social
insurance	school	trade	security
tax	federal	chemicals	insurance
congress	aid	tariff	medical
income	government	congress	care
minimum	tax	drugs	medicare
wage	energy	communicable	disability
business	research	diseases	assistance

s_j corresponds to topic j. $G(s, i)$ is the group of s on topic i which has a size of $|G(s, i)|$ and $G(s, j)$ is the group of s on topic j. We present entities that frequently change their group alliance according to the topics of resolutions.

Group cohesion from the GT model is found to be significantly greater than the baseline group cohesion under a pairwise t-test, as shown in Table 2, which indicates that the GT's joint inference is better able to discover cohesive groups. We find that nearly every document has a higher Agreement Index across groups using the GT model as compared to the baseline. As expected, stochastic block structures without text [9] is even worse than our baseline.

4.1 The US Senate Dataset

Our Senate dataset consists of the voting records of Senators in the 101st-109th US Senate (1989-2005) obtained from the Library of Congress THOMAS database. During a roll call for a particular bill, a Senator may respond *Yea* or *Nay* to the question that has been put to vote, else the vote will be recorded as *Not Voting*. We do not consider *Not Voting* as a unique vote since most of the time it is a result of a Senator being absent from the session of the US Senate.

Table 5. Senators in the four groups corresponding to Topic Education + Domestic in Table 4

Group 1	Group 3	Group 4
73 Republicans	Cohen(R-ME)	Armstrong(R-CO)
Krueger(D-TX)	Danforth(R-MO)	Garn(R-UT)
Group 2	Durenberger(R-MN)	Humphrey(R-NH)
90 Democrats	Hatfield(R-OR)	McCain(R-AZ)
Chafee(R-RI)	Heinz(R-PA)	McClure(R-ID)
Jeffords(I-VT)	Kassebaum(R-KS)	Roth(R-DE)
	Packwood(R-OR)	Symms(R-ID)
	Specter(R-PA)	Wallop(R-WY)
	Snowe(R-ME)	Brown(R-CO)
	Collins(R-ME)	DeWine(R-OH)
		Thompson(R-TN)
		Fitzgerald(R-IL)
		Voinovich(R-OH)
		Miller(D-GA)
		Coleman(R-MN)

Table 6. Senators that switch groups the most across topics for the 101st-109th Senates

Senator	Group Switch Index
Shelby(D-AL)	0.6182
Heflin(D-AL)	0.6049
Voinovich(R-OH)	0.6012
Johnston(D-LA)	0.5878
Armstrong(R-CO)	0.5747

The text associated with each resolution is composed of its index terms provided in the database. There are 3423 resolutions in our experiments (we excluded roll calls that were not associated with resolutions). Each bill may come up for vote many times in the U.S. Senate, each time with an attached amendment, and thus many relations may have the same attributes (index terms). Since there are far fewer words than pairs of votes, we adjust the text likelihood to the 5th power (weighting factor 5) in the experiments with this dataset so as to balance its influence during inference.

We cluster the data into 4 topics and 4 groups (cluster sizes are suggested by a political science professor) and compare the results of GT with the baseline. The most likely words for each topic from the traditional mixture of unigrams model is shown in Table 3, whereas the topics obtained using GT are shown in Table 4. The GT model collapses the topics Education and Energy together into Education and Domestic, since the voting patterns on those topics are quite similar. The new topic Social Security + Medicare did not have strong enough word coherence to appear in the baseline model, but it has a very distinct voting pattern, and thus is clearly found by the GT model. Thus GT discovers topics

Table 7. Top words for topics generated from mixture of unigrams model with the UN dataset (1990-2003). Only text information is utilized to form the topics, as opposed to Table 8 where our GT model takes advantage of both text and voting information.

Everything Nuclear	Human Rights	Security in Middle East
nuclear	rights	occupied
weapons	human	israel
use	palestine	syria
implementation	situation	security
countries	israel	calls

that are salient in that they correlate with people's behavior and relations, not simply word co-occurrences.

Examining the group distribution across topics in the GT model, we find that on the topic Economic the Republicans form a single group whereas the Democrats split into 3 groups indicating that Democrats have been somewhat divided on this topic. With regard to Education + Domestic and Social Security + Medicare, Democrats are more unified whereas the Republicans split into 3 groups. The group membership of Senators on Education + Domestic issues is shown in Table 5. We see that the first group of Republicans include a Democratic Senator from Texas, a state that usually votes Republican. Group 2 (majority Democrats) includes Sen. Chafee who is known to be pro-environment and is involved in initiatives to improve education, as well as Sen. Jeffords who left the Republican Party to become an Independent and has championed legislation to strengthen education and environmental protection.

Nearly all the Senators in Group 4 (in Table 5) are advocates for education and many of them have been awarded for their efforts (e.g., Sen. Fitzgerald has been honored by the NACCP for his active role in Early Care and Education, and Sen. McCain has been added to the ASEE list as a *True Hero* in American Education). Sen. Armstrong was a member of the Education committee; Sen. Voinovich and Sen. Symms are strong supporters of early education and vocational education, respectively; and Sen. Roth has constantly voted for tax deductions for education. It is also interesting to see that Sen. Miller (D-GA) appears in a Republican group; although he is in favor of educational reforms, he is a conservative Democrat and frequently criticizes his own party—even backing Republican George W. Bush over Democrat John Kerry in the 2004 Presidential election.

Many of the Senators in Group 3 have also focused on education and other domestic issues such as energy, however, they often have a more liberal stance than those in Group 4, and come from states that are historically less conservative. Senators Hatfield, Heinz, Snowe, Collins, Cohen and others have constantly promoted pro-environment energy options with a focus on renewable energy, while Sen. Danforth has presented bills for a more fair distribution of energy resources. Sen. Kassebaum is known to be uncomfortable with many Republican views on

Table 8. Top words for topics generated from the GT model with the UN dataset (1990-2003) as well as the corresponding groups for each topic (column). The countries listed for each group are ordered by their 2005 GDP (PPP) and only the top 5 countries are shown in groups that have more than 5 members.

G	Nuclear Arsenal	Human Rights	Nuclear Arms Race
R	nuclear	rights	nuclear
O	states	human	arms
U	united	palestine	prevention
P	weapons	occupied	race
↓	nations	israel	space
1	Brazil	Brazil	UK
	Columbia	Mexico	France
	Chile	Columbia	Spain
	Peru	Chile	Monaco
	Venezuela	Peru	East-Timor
2	USA	Nicaragua	India
	Japan	Papua	Russia
	Germany	Rwanda	Micronesia
	UK...	Swaziland	
	Russia	Fiji	
3	China	USA	Japan
	India	Japan	Germany
	Mexico	Germany	Italy...
	Iran	UK...	Poland
	Pakistan	Russia	Hungary
4	Kazakhstan	China	China
	Belarus	India	Brazil
	Yugoslavia	Indonesia	Mexico
	Azerbaijan	Thailand	Indonesia
	Cyprus	Philippines	Iran
5	Thailand	Belarus	USA
	Philippines	Turkmenistan	Israel
	Malaysia	Azerbaijan	Palau
	Nigeria	Uruguay	
	Tunisia	Kyrgyzstan	

domestic issues such as education, and has voted against voluntary prayer in school. Thus, both Groups 3 and 4 differ from the Republican core (Group 2) on domestic issues, and also differ from each other.

The Senators that switch groups the most across topics in the GT model are shown in Table 6 based on their GSIs. Sen. Shelby(D-AL) votes with the Republicans on Economic, with the Democrats on Education + Domestic and with a small group of maverick Republicans on Foreign and Social Security + Medicare. Both Sen. Shelby and Sen. Heflin are Democrats from a fairly conservative state (Alabama) and are found to side with the Republicans on many issues.

4.2   The United Nations Dataset

The second dataset involves the voting record of the UN General Assembly [21]. We focus first on the resolutions discussed from 1990-2003, which contain votes of 192 countries on 931 resolutions. If a country is present during the roll call, it may choose to vote *Yes*, *No* or *Abstain*. Unlike the Senate dataset, a country's vote can have one of three possible values instead of two. Because we parameterize agreement and not the votes themselves, this 3-value setting does not require any change to our model. In experiments with this dataset, we use a weighting factor 500 for text (adjusting the likelihood of text by a power of 500 so as to make it comparable with the likelihood of pairs of votes for each resolution). We cluster this dataset into 3 topics and 5 groups (again, numbers are suggested by a political science professor).

The most probable words in each topic from the mixture of unigrams model is shown in Table 7. For example, Everything Nuclear constitutes all resolutions that have anything to do with the use of nuclear technology, including nuclear weapons. Comparing these with topics generated from the GT model shown in Table 8, we see that the GT model splits the discussion about nuclear technology into two separate topics, Nuclear Arsenal which is generally about countries obtaining nuclear weapons and management of nuclear waste, and Nuclear Arms Race which focuses on the arms race between Russia and the US and preventing a nuclear arms race in outer space. These two issues had drastically different voting patterns in the U.N., as can be seen in the contrasting group structure for those topics in Table 8. The countries in Table 8 are ranked by their GDP in 2005.[1] Thus, again the GT model is able to discover salient topics—topics that reflect the voting patterns and coalitions, not simply word co-occurrence alone.

As seen in Table 8, groups formed in Nuclear Arms Race are unlike the groups formed in the remaining topics. These groups map well to the global political situation of that time when, despite the end of the Cold War, there was mutual distrust between Russia and the US with regard to the continued manufacture of nuclear weapons. For missions to outer space and nuclear arms, India was a staunch ally of Russia, while Israel was an ally of the US.

Overlapping Time Intervals. In order to understand changes and trends in topics and groups over time, we run the GT model on resolutions that were discussed during overlapping time windows of 15 years, from 1960-2000, each shifted by a period of 5 years. We consider 3823 unique resolutions in this way. The topics as well as the group distribution for the most dominant topic during each time period are shown in Table 9.

Over the years there is a shift in the topics discussed in the UN, which corresponds well to the events and issues in history. During 1960-1975 the resolutions focused on countries having the right to self-determination, especially countries in Africa which started to gain their freedom during this time. Although this

---

[1] http://en.wikipedia.org/wiki/List_of_countries_by_GDP_%28PPP%29. In Table 8, we omit some countries (represented by ...) in order to incorporate other interesting but relatively low ranked countries (for example, Russia) in the GDP list.

Table 9. Results for 15-year-span slices of the UN dataset (1960-2000). The top probable words are listed for all topics, but only the groups corresponding the most dominant topic are shown (Topic 3). We list the countries for each group ordered by their 2005 GDP (PPP)and only show the top 5 countries in groups that have more than 5 members. We do not repeat the results in Table 8 for the most recent window (1990-2003).

Time Period	Topic 1	Topic 2	Topic 3	Group distributions for Topic 3				
				Group 1	Group2	Group3	Group4	Group5
60-75	Nuclear	Procedure	Africa Indep.	India	USA	Argentina	USSR	Turkey
	operative	committee	calling	Indonesia	Japan	Colombia	Poland	
	general	amendment	right	Iran	UK	Chile	Hungary	
	nuclear	assembly	africa	Thailand	France	Venezuela	Bulgaria	
	power	deciding	self	Philippines	Italy	Dominican	Belarus	
65-80	Independence	Finance	Weapons	Cuba	India	Algeria	USSR	USA
	territories	budget	nuclear	Albania	Indonesia	Iraq	Poland	Japan
	independence	appropriation	UN		Pakistan	Syria	Hungary	UK
	self	contribution	international		Saudi	Libya	Bulgaria	France
	colonial	income	weapons		Egypt	Afghanistan	Belarus	Italy
70-85	N. Weapons	Israel	Rights	Mexico	China	USA	Brazil	India
	nuclear	israel	africa	Indonesia		Japan	Argentina	USSR
	international	measures	territories	Iran		UK	Colombia	Poland
	UN	hebron	south	Thailand		France	Chile	Vietnam
	human	expelling	right	Philippines		Italy		Hungary
75-90	Rights	Israel/Pal.	Disarmament	Mexico	USA	Algeria	China	India
	south	israel	UN	Indonesia	Japan	Vietnam	Brazil	
	africa	arab	international	Iran	UK	Iraq	Argentina	
	israel	occupied	nuclear	Thailand	France	Syria	Colombia	
	rights	palestine	disarmament	Philippines	USSR	Libya	Chile	
80-95	Disarmament	Conflict	Pal. Rights	USA	China	Japan	Guatemala	Malawi
	nuclear	need	rights	Israel	India	UK	St Vincent	
	US	israel	palestine		Russia	France	Dominican	
	disarmament	palestine	israel		Spain	Italy		
	international	secretary	occupied		Hungary	Canada		
85-00	Weapons	Rights	Israel/Pal.	Poland	China	USA	Russia	Cameroon
	nuclear	rights	israeli	Czech R.	India	Japan	Argentina	Congo
	weapons	human	palestine	Hungary	Brazil	UK	Ukraine	Ivory C.
	use	fundamental	occupied	Bulgaria	Mexico	France	Belarus	Liberia
	international	freedoms	disarmament	Albania	Indonesia	Italy	Malta	

topic continued to be discussed in the subsequent time period, the focus of the resolutions shifted to the role of the UN in controlling nuclear weapons as the Cold War conflict gained momentum in the late 70s. While there were few resolutions condemning the racist regime in South Africa between 1965-1980, this was the topic of many resolutions during 1970-1985—culminating in the UN censure of South Africa for its discriminatory practices.

Other topics discussed during the 70s and early 80s were Israel's occupation of neighboring countries and nuclear issues. The reduction of arms was primarily discussed during 1975-1990, the time period during which the US and Soviet Union had talks about disarmament. During 1980-1995 the central topic of discussion was the Israeli-Palestinian conflict; this time period includes the beginning of the *Intifada* revolt in Palestine and the Gulf War. This topic continued to be important in the next time period (1985-2000), but in the most recent slice (1990-2003, Table 8) it has become a part of a broader topic on human rights by combining other human rights related resolutions that appear as a separate topic during 1985-2000. The human rights issue continues to be the primary topic of discussion during 1990-2003.

Throughout the history of the UN, the US is usually in the same group as Europe and Japan. However, as we can see in Table 9 during 1985-2000, when the Israeli-Palestinian conflict was the most dominant topic, US and Israel form a group of their own separating themselves from Europe. In other topics discussed during 1985-2000, US and Israel are found to be in the same group as Europe and Japan.

Another interesting result of considering the groups formed over the years is that, except for the last time period (1990-2003), countries in eastern Europe such as Poland, Hungary, Bulgaria, etc., form a group along with USSR (Russia). However, in the last time window on most topics they become a part of the group that consists of the western Europe, Japan and the US. This shift corresponds to the end of the communist regimes in these countries that were supported by the Soviet Union. It is also worth mentioning that before 1990, our model assigned East Germany to the same group as other eastern European countries and USSR (Russia), while it assigned West Germany to the same group as western European countries.[2]

5   Groups over Time

In Table 9 in the previous section, we show that group formation changes over time by simply pre-dividing the dataset into disjoint subsets based on timestamps. In this section, by contrast, we investigate the dynamic changes of groups using a model that explicitly incorporates time—jointly discovering groups and their continuous time profiles.

Traditional transition-based Markov models have played a major role in modeling various dynamic systems including social networks. For example, recent work by Sarkar and Moore [22] proposes a latent space model that accounts for friendships drifting over time. Blei and Lafferty propose dynamic topic models (DTMs) in which the alignment among topics across time steps is captured by a Kalman filter [23].

Instead we propose here a new model that does not make the Markov assumption, rather, we treat timestamps as observed random variables, as in [24]. In comparison to more complex alternatives, the relative simplicity of our model is a great advantage—not only for the relative ease of understanding and implementing it, but also because this approach can in the future be naturally applied into other more richly structured models. In our model, which we call *Groups over Time (GOT)*, group discovery is influenced not only by relational co-occurrences, but also by temporal information. Rather than modeling a sequence of state changes with a Markov assumption on the dynamics, GOT models (normalized) absolute timestamp values. This allows GOT to see long-range dependencies in time, and to predict group distributions given a timestamp. It also helps avoid a Markov model's risk of inappropriately dividing a group in two when there is a brief gap in its appearance.

The graphical model representation of our model is shown in Figure 2. For comparison, the stochastic block structures model is shown in Figure 2(a). Groups over Time is a generative model of timestamps and the relational structures of a social network. There are two ways of describing its generative process. The first, which corresponds to the process used in Gibbs sampling for parameter estimation (Figure 2(c)), is as follows:

---

[2] This is not shown in Table 9 because they are missing from the 2005 GDP data.

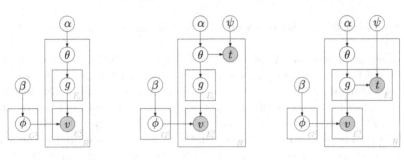

(a) Multi-relation
stochastic block structures

(b) GOT model,
alternate view

(c) GOT model,
for Gibbs sampling

Fig. 2. Three group models: stochastic block structures and two perspectives on GOT

1. Draw G^2 binomials ϕ_{gh} from a Dirichlet prior β, one for each group pair (g, h);
2. For each relation r in total R relations, draw a multinomial θ_r from a Dirichlet prior α;
   (a) For each entity e_{ri} of E_r entities in relation r, draw a group g_{ri} from multinomial θ_r and draw a timestamp t_{ri} from Beta $\psi_{g_{ri}}.$;
   (b) For each entity pair (i, j), draw their agreement v_{ij} from binomial $\phi_{g_i g_j}$;

Although, in the above generative process, a timestamp is generated for each entity, all the timestamps of the entities in a relation are observed as the same, because there is typically only one timestamp associated with a relation. When fitting our model from typical data, each training relation's timestamp is copied to all the entities appearing in the relation. However, after fitting, if actually run as a generative model, this process would generate different time stamps for the entities appearing in the same relation. An alternative generative process description of GOT, (better suited to generate an unseen relation), is one in which a single timestamp is associated with each relation, generated by rejection or importance sampling, from a mixture of per-group Beta distributions over time with mixtures weight as the per-relation θ_r over groups. As before, this distribution over time would be parameterized by the set of timestamp-generating Beta distributions, one per group. The graphical model for this alternative generative process is shown in Figure 2(b).

5.1 Dynamic Group Discovery in UN

We apply the Group over Time (GOT) model to the UN data set described in Section 4.2, and compare it with the stochastic block structures model. Because of space limitation, we do not show example group distributions from the two models. However, when we calculate the Agreement Index (AI, defined in Section 4) of the groups discovered by the two models on the UN data set; we find that the average AI for the stochastic block structures model is 0.7934, and 0.8169 for GOT. We conclude that the groups obtained from the GOT model are significantly more cohesive (p-value $< .01$) than those obtained from the block structures model.

Note that the average AI from GOT (0.8169) is also lower than the one from GT (0.8664) due to the lack of textual attributes.

6   Conclusions

We present the Group-Topic model that jointly discovers latent groups in a network as well as clusters of attributes (or topics) of events that influence the interaction between entities in the network. The model extends prior work on latent group discovery by capturing not only pair-wise relations between entities but also multiple attributes of the relations (in particular, the model considers words describing the relations). In this way the GT model obtains more cohesive groups as well as fresh topics that influence the interaction between groups. The model could be applied to variables of other data types in addition to voting data. We are now using the model to analyze the citations in academic papers to capture the topics of research papers and discover research groups. It would also apply to a much larger network of entities (people, organizations, etc.) that frequently appear in newswire articles.

The model can be altered suitably to consider other attributes characterizing relations between entities in a network. In ongoing work we are extending the Group-Topic model to capture a richer notion of topic, where the attributes describing the relations between entities are represented by a mixture of topics.

The Group over Time model provides a simple and novel way to take advantage of the temporal information as continuous observations, in contrast to the traditional transition-based Markov models. We believe that the simplicity of this approach is an advantage in certain applications.

Acknowledgments

This work was supported in part by the National Science Foundation under grant #IIS-0326249, and by the Defense Advanced Research Projects Agency under contract #NBCHD030010 and #HR0011-06-C-0023. We would also like to greatly thank Prof. Vincent Moscardelli, Chris Pal and Aron Culotta for helpful discussion.

References

1. Wasserman, S., Faust, K.: Social Network Analysis: Methods and Applications. Cambridge University Press (1994)
2. Carley, K.: A comparison of artificial and human organizations. Journal of Economic Behavior and Organization **56** (1996) 175–191
3. Denham, W.W., McDaniel, C.K., Atkins, J.R.: Aranda and Alyawarra kinship: A quantitative argument for a double helix model. American Ethnologist **6**(1) (1979) 1–24
4. Sparrow, M.: The application of network analysis to criminal intelligence: an assessment of prospects. Social Networks **13** (1991) 251–274
5. Hix, S., Noury, A., Roland, G.: Power to the parties: Cohesion and competition in the European Parliament, 1979-2001. British Journal of Political Science **35**(2) (2005) 209–234

6. Cox, G., Poole, K.: On measuring the partisanship in roll-call voting: The U.S. House of Represenatatives, 1887-1999. American Journal of Political Science **46**(1) (2002) 477–489

7. Kubica, J., Moore, A., Schneider, J., Yang, Y.: Stochastic link and group detection. In: Proceedings of the 17th National Conference on Artificial Intelligence. (2002)

8. Bhattacharya, I., Getoor, L.: Deduplication and group detection using links. In: Workshop on Link Analysis and Group Detection, in conjunction with the 10th International ACM SIGKDD Conference. (2004)

9. Nowicki, K., Snijders, T.A.: Estimation and prediction for stochastic blockstructures. Journal of the American Statistical Association **96**(455) (2001)

10. Kemp, C., Griffiths, T.L., Tenenbaum, J.: Discovering latent classes in relational data. Technical report, MIT CSAIL (2004)

11. McCallum, A., Corrada-Emanuel, A., Wang, X.: Topic and role discovery in social networks. In: Proceedings of the 19th International Joint Conference on Artificial Intelligence. (2005)

12. Blei, D., Ng, A., Jordan, M.: Latent Dirichlet allocation. Journal of Machine Learning Research **3** (2003) 993–1022

13. Erosheva, E.A., Fienberg, S.E., Lafferty, J.: Mixed-membership models of scientific publications. Proceedings of the National Academy of Sciences **97**(22) (2004) 11885–11892

14. Airoldi, E.M., Blei, D.M., Xing, E.P., Fienberg, S.E.: A latent mixed-membership model for relational data. In: Workshop on Link Discovery: Issues, Approaches and Applications, in conjunction with the 11th International ACM SIGKDD Conference. (2005)

15. Dhillon, I.S., Mallela, S., Modha, D.S.: Information-theoretic co-clustering. In: Proceedings of the 9th ACM SIGKDD International Conference on Knowledge Discovery and Data Mining. (2003)

16. Bekkerman, R., Yaniv, R.E., McCallum, A.: Multi-way distributional clustering via pairwise interactions. In: Proceedings of the 22nd International Conference on Machine Learning. (2005)

17. Jakulin, A., Buntine, W.: Analyzing the US Senate in 2003: Similarities, networks, clusters and blocs (2004)

18. Pajala, A., Jakulin, A., Buntine, W.: Parliamentary group and individual voting behavior in Finnish Parliamentin year 2003: A group cohesion and voting similarity analysis (2004)

19. Beeferman, D., Berger, A.: Agglomerative clustering of a search engine query log. In: Proceedings of the 6th ACM SIGKDD International Conference on Knowledge Discovery and Data Mining. (2000)

20. Fenn, D., Suleman, O., Efstathiou, J., Johnson, N.: How does Europe make its mind up? Connections, cliques, and compatibility between countries in the Eurovision song contest. arXiv:physics/0505071 (2005)

21. Voeten, E.: Documenting votes in the UN General Assembly (2003) http://home.gwu.edu/~voeten/UNVoting.htm.

22. Sarkar, P., Moore, A.: Dynamic social network analysis using latent space models. In: Advances in Neural Information Processing Systems 18. (2005)

23. Blei, D.M., Lafferty, J.D.: Dynamic topic models. In: Proceedings of the 23rd International Conference on Machine Learning. (2006)

24. Wang, X., McCallum, A.: Topics over time: A non-markov continuous-time model of topical trends. In: Proceedings of the 12th ACM SIGKDD International Conference on Knowledge Discovery and Data Mining. (2006)

Statistical Models for Networks:
A Brief Review of Some Recent Research

Stanley Wasserman[1], Garry Robins[2], and Douglas Steinley[3]

[1] Department of Statistics, Department of Sociology, and Department of
Psychological and Brain Sciences, Indiana University
stanwass@indiana.edu
[2] School of Behavioural Science, University of Melbourne
g.robins@psych.unimelb.edu.au
[3] Department of Psychology and Department of Statistics, University of Missouri
steinleyd@missouri.edu

We begin with a graph (or a directed graph), a single set of nodes $\mathcal{N}$, and a set
of lines or arcs $\mathcal{L}$. It is common to use this mathematical concept to represent
a *network*. We use the notation of [1], especially Chapters 13 and 15. There
are extensions of these ideas to a wide range of networks, including multiple
relations, affiliation relations, valued relations, and social influence and selection
situations (in which information on attributes of the nodes is available), all of
which can be found in the chapters of [2].

The purpose of this short exposition is to discuss the developments in statistical models for networks that have occurred over the past five years, since the
publication of the statistical chapters (8, 9, 10, and 11) of Carrington, Scott,
and Wasserman (which were written in 2002). The statistical modeling of social networks is advancing quite quickly. The many exciting new developments
include, for instance, longitudinal models for the co-evolution of networks and
behavior [3] and latent space models for social networks [4]. In this chapter, we
do not intend to review *all* the recent advances but rather limit our scope to a
few developments that we have worked on.

1   Background

Early work on distributions for graphs was quite limiting, forcing researchers to
adopt independence assumptions that were not terribly realistic (see Chapters
13-16 of [1]. It is hard to accept the standard assumption common in much
of the literature, especially in physics, of complete independence and then to
adopt the mis-named and overly simplistic "random graph" distribution (there
are, of course, an infinite number of random graph distributions). The random
graph distribution to the physicists, that is usually referred to as a Bernoulli
graph, assumes no dependencies at all among the random components of a graph.
Equally hard to believe as a true representation of social behavior are the many
conditional uniform distributions and p_1, which assumes independent dyads.

The breakthrough in statistical modeling of networks was first exposited by
[5], who termed their model a *Markov random graph*. Further developments,

E.M. Airoldi et al. (Eds.): ICML 2006 Ws, LNCS 4503, pp. 45–56, 2007.

especially commentary on estimation of distribution parameters, were given by [6]. [7] elaborated upon the model, describing a more general family of distributions. [8, 9] and [10] further developed this family of models, showing how a Markov parametric assumption gives just one, of many, possible sets of parameters. This family, with its variety and extensions, was named p^*, a label by which it has come to be known. The parameters (which are determined by the hypothesized dependence structure) reflect structural concerns, which are assumed to be governing the probabilistic nature of the underlying social and/or behavioral process.

This pre-2000 early work by the first researchers extended p^* in a variety of ways, and laid the foundation for work in this decade on the estimation problems inherent in the early formulations. This research also was an important forerunner of the new parametric specifications that promise wider usage of the family. A more thorough history of this family of distributions, including a discussion of its roots in spatial modeling and statistical physics, can be found in [11]. [12] offers a review of p^* circa-2003, while [13] reviews the 2003-2006 period.

The work of [5] did indeed begin a new era for statistical modeling of networks, although it took ten years for Markov random graphs to be discussed at more length by network methodologists. We briefly describe the highlights of the past decade here.

2   Some Notation and a Bit of Background

A *network* is a set of n actors and a collection of r relations that specify how these actors are related to one another. As defined by [1] (Chapter 3), a network can also contain a collection of attribute characteristics, measured on the actors.

We let $\mathcal{N} = \{1, 2, \ldots, g\}$ denote the set of actors, and $\mathcal{X}$ denote a particular relation defined on the actors (here, we let $r = 1$). Specifically, $\mathcal{X}$ is a set of ordered pairs recording the presence or absence of relational ties between pairs of actors. This binary relation can be represented by a $g \times g$ matrix $\mathbf{X}$, with elements

$$X_{ij} = \begin{cases} 1 \text{ if } (i, j) \in \mathcal{X}, \\ 0 \text{ otherwise.} \end{cases}$$

We will use a variety of graph characteristics and statistics throughout; such quantities are defined in the early chapters of [1]. We assume throughout that $\mathbf{X}$ and its elements are random variables. Typically, these variables are assumed to be interdependent, given the interactive nature of the social processes that generate and sustain a network. Much of the work over the past decade has been on the explicit hypotheses underlying different types of interpendencies.

In fact, one of the new ideas for network analysis, utilized by the p^* family of models is a *dependence graph*, a device which allows one to consider which elements of $\mathbf{X}$ are independent. [12] discusses such graphs at length. A dependence graph, which we illustrate in the next section, is also the starting point for the Hammersley–Clifford Theorem, which posits a very general probability distribution for network random variables using the postulated dependence graph. The exact form of the dependence graph depends on the nature of the substantive hypotheses about the network under study.

As outlined by [13], a statistical model for a network can be constructed using this approach through a series of five steps:

- Regard each relational tie as a random variable
- Specify a dependence hypothesis, leading to a dependence graph
- Generate a specific model from the p^* family from the specified dependence graph
- Simplify the parameter space through homogeneity or other constraints
- Estimate, assess, and interpret model parameters

We have mentioned the first two of these steps. To discuss the latter three, we need to introduce p^* via a dependence graph.

3 Statistical Theory

Any observed single relational network may be regarded as a realization $\mathbf{x} = [x_{ij}]$ of a random two-way binary array $\mathbf{X} = [X_{ij}]$. The dependence structure for these random variables is determined by the *dependence graph* $\mathcal{D}$ of the random array $\mathbf{X}$. $\mathcal{D}$ is itself a graph whose nodes are elements of the index set $\{(i, j); i, j \in \mathcal{N}, i \neq j\}$ for the random variables in $\mathbf{X}$, and whose edges signify pairs of the random variables that are assumed to be conditionally dependent (given the values of all other random variables).

More formally, a dependence graph for a univariate network has node set

$$\mathcal{N}_D = \{(i, j); i, j \in \mathcal{N}, i \neq j\}.$$

The edges of $\mathcal{D}$ are given by

$$\mathcal{E}_D = \{((i, j), (k, l)), \text{ where } X_{ij} \text{ and } X_{kl} \text{ are not conditionally independent}\}.$$

Consider now a general dependence graph, with an arbitrary edge set. Such a dependence graph yields a very general probability distribution for a (di)graph, which we term p^* and focus on below.

For an observed network, which we consider to be a realization $\mathbf{x}$ of a random array $\mathbf{X}$, we assume the existence of a dependence graph $\mathcal{D}$ for the random array $\mathbf{X}$. The edges of $\mathcal{D}$ are crucial here; consider the set of edges, and determine if there are any *complete subgraphs*, or cliques found in the dependence graph. (For a general dependence graph, a subset A of the set of relational ties $\mathcal{N}_D$ is *complete* if every pair of nodes in A (that is, every pair of relational ties) is linked by an edge of $\mathcal{D}$. A subset comprising a single node is also regarded as complete.). These cliques specify which subsets of relational ties are all pairwise, conditionally dependent on each other.

The Hammersley–Clifford theorem (see [12] for a summary) establishes that a probability model for $\mathbf{X}$ depends only on the cliques of the dependence graph $\mathcal{D}$.

In particular, application of the Hammersley–Clifford theorem yields a characterization of $Pr(\mathbf{X} = \mathbf{x})$ in the form of an exponential family of distributions:

$$Pr(\mathbf{X} = \mathbf{x}) = \left(\frac{1}{\kappa}\right) \exp\left(\sum_{A \subseteq \mathcal{N}_D} \lambda_A \prod_{(i,j) \in A} x_{ij}\right) \tag{1}$$

where:

- $\kappa = \sum_{\mathbf{x}} \exp\{\sum_{A \subseteq \mathcal{D}} \lambda_A \prod_{(i,j) \in A} x_{ij}\}$ is a normalizing quantity;
- $\mathcal{D}$ is the dependence graph for $\mathbf{X}$; the summation is over all subsets A of nodes of $\mathcal{D}$;
- $\prod_{(i,j) \in A} x_{ij}$ is the sufficient statistic corresponding to the parameter λ_A; and
- $\lambda_A = 0$ whenever the subgraph induced by the nodes in A is not a clique of $\mathcal{D}$.

The set of non–zero parameters in this probability distribution for $Pr(\mathbf{X} = \mathbf{x})$ depends on the *maximal* cliques of the dependence graph (A maximal clique is a complete subgraph that is not contained in any other complete subgraph). Any subgraph of a complete subgraph is also complete (but not maximal), so that if A is a maximal clique of $\mathcal{D}$, then the probability distribution for the (di)graph will contain non–zero parameters for A and all of its subgraphs. Each clique, and hence each nonzero parameter in the model, corresponds to a *configuration*, a small subgraph of possible network ties. Different dependence assumptions result in different types of configurations. For instance, [5] showed that configurations for Markov dependence (described below) were edges, stars of various types (a single node with arcs going in and/or out), and triangles for nondirected graphs. The model in effect supposes that the observed network is built up from combinations of these various configurations, and the parameters express the presence (or absence) of the configurations in the observed network. For instance, a strongly positive triangle parameter is evidence for more triangulation in the network implying that networks with large numbers of triads have larger probabilities of arising.

All models from this family, which we refer to as p^*, have this form. Some recent literature refers to these models as ERGMs — exponential random graph models. It is our course uninformative to refer to these distributions as "exponential random graphs" — almost any probability distribution for a graph can be made "exponential". Further, strictly speaking, the model is not exponential, but in the statistical sense, an exponential *family*, which conveys a special meaning in statistical theory (and has important implications for some of the estimation procedures described below – see [14]. Hence, we much prefer the more informative moniker p^*, and the label, *an exponential family of distributions* for random graphs. The p^* label (first used by [7] derives from the research on statistical modeling commenced by Holland and Leinhardt with their dyadic–independence p_1 model.

As for the details: the probability of a particular realization of a random graph depends on the cliques of the dependence graph, and from that, the sufficient statistics (arising from the *configurations*) specified by the hypothesized dependencies. The sufficient statistics are the counts of these configurations arising in the realization being modeled.

There are a variety of dependence graphs well-known in the literature. One very general and simple member of the p^* family is the Bernoulli graph; another is Holland and Leinhardt's p_1. The dependence graph first proposed for networks, for which this distribution was first developed, assumes conditional independence of X_{ij} and X_{kl} if and only if $\{i, j\} \cap \{k, l\} = \emptyset$. This dependence graph links any two relational ties involving the same actor(s); thus, any two relational ties are associated if they involve the same actor(s). Because of the similarity to the dependence inherent in a Markov spatial process, such a random graph was labeled "Markov" by [5]. Further discussion can be found in Section 4 of [13].

One can also formulate dependence graphs when data on attribute variables measured on the nodes is available. If the attribute variables are taken as fixed, with network ties varying depending on the attributes, then models for social selection arise [15]. If on the other hand the network is assumed fixed, with the distribution of attributes dependent on the pattern of network ties, the outcomes are models for social influence [15].

4 Parameters – New Specifications

Limiting the number of parameters is wise – one can either postulate a simple dependence graph, or by making assumptions about the parameters. The usual assumption is *homogeneity*, in which parameters for isomorphic *configurations* of nodes are equated.

Even with homogeneity imposed, models may not be identifiable. Typically, parameters for higher order configurations (for example, higher order stars or triads) are set to zero (equivalent to setting higher order interactions to zero in general linear models).

As mentioned, Markov random graph models were indeed a breakthrough in moving towards more realistic dependence assumptions. But recently it has been shown that Markov dependence is often inadequate in handling typical social network data. Frequently, parameters arising from Markov dependence assumptions are consistent with either complete or very sparse networks, which are of course unhelpful in modeling realistic data. Several authors have provided technical demonstrations of this problem [16, 17, 18, 19, 20, 21]. An intuitive explanation of the difficulty follows.

Markov random graphs assume that stars and triangles are rather evenly spread throughout the network. In many observed networks, of course, there are dense regions (that is, concentrations of many triangles) and some high degree nodes (that is, concentrations of many stars). As a result, for such data, parameter estimation for Markov random graphs is problematic: there is difficulty in finding "average" parameter values that can adequately capture such structural heterogeneity (see [19] and [21], for further discussion.) When sensible parameter estimates cannot be obtained, the model is said to be *degenerate*, or *nonconvergent*.

[21] proposed a method of combining counts of all the Markov star parameters into the one statistic, with geometrically decreasing weights on the higher order

star counts so that they did not come to dominate the calculation. The resulting parameter is termed a *geometrically weighted degree parameter*, or an *alternating k-star parameter* (the term alternating comes from alternating signs in the calculation of the statistic.) Various versions of this new degree-based parameter have been proposed (see [14], who shows the linkages between them), but whatever the precise form of the parameter, it permits greater heterogeneity in the degree distribution, so that it is more capable of modeling high degree nodes than a small number of low order Markov star parameters. Such parameters appear to greatly increase the "fittability" of models.

Perhaps the most important innovation of [21] was the introduction of *k-triangles*, configurations with k separate triangles sharing one edge, the base of the k-triangle. These configurations also introduce a new distribution of graph features (alongside the degree distribution and the geodesic distribution): the *edge-wise shared partner distribution* (see [14, 22]). Counts of the k-triangle configurations are combined into one statistic just as for the case of the geometrically weighted degree parameter, producing a new statistic and associated parameter for alternating k-triangles. This parameter models triangulation in the network but permits more heterogeneity. Alternating k-triangles are much better than the Markov single triangle parameter in dealing with clumps of triangles that form the denser regions of the network. The parameter has a simple general interpretation: a large positive parameter value indicates that there is substantial triangulation in the network, and that this is likely to be expressed in the formation of denser regions.

[21] also proposed k-paths, configurations identical to k-triangles except that the edge at the base of the k-triangle is not necessarily present. This configuration quantifies multiple independent paths between pairs of nodes. Again, [21] combined these configurations into the one parameter, alternating k-paths. There is an associated distribution across the graph, the *dyad-wise shared partner distribution* (that is, shared partners based on dyads, not just on edges; see [14, 22].

As the new degree parameter is based on a combination of Markov star configurations, it can be derived from Markov dependence. Markov dependence alone, however, is not sufficient to produce the alternating k-triangle and k-path parameters, which require higher order dependence structures [?]. The additional dependence assumption is referred to by [19] as *social circuit dependence*, where the presence of two edges in the observed graph creates dependence among two other possible edges, assuming the four edges constitute a 4-cycle. Social circuit dependence appears complicated but it reflects a simple feature of social interaction. For example, if John and Mary work together, and if Joanne and Mark work together, then a working relationship between John and Joanne may increase the chances of a working relationship between Mary and Mark. But the argument simply does not work without the existing John/Mary and Joanne/Mark working relationships. In other words, the ties between John and Mary, and Joanne and Mark, create the dependence between possible John/Joanne and Mary/Mark ties.

This is a special, and rather different, dependence assumption. First, it explicitly permits the emergence of dependence through existing observations, in the sense that the presence of certain ties creates dependencies that otherwise could not exist. Secondly, this emergent dependence permits the appearance of higher-order structures (for example, "clumps" of triangles). Thirdly, the necessity of this assumption gives evidence to the importance of emergent processes in networks. It is a *self-organizing* quality, apparent in real social networks. Self-organizing systems imply that the presence or absence of certain ties may lead to the creation, maintenance, or disappearance of other ties. While the simpler Markov models can also be interpreted in terms of such self-organizing qualities, the social circuit dependence enables the appearance of higher order structures (e.g. dense regions of triangles) that are expressly implied by the model, rather than a simple chance accumulation of basic Markov configurations. Social networks are complex systems that typically cannot be adequately represented with simplified assumptions.

We note that these new specifications do not resolve all the problems of degeneracy and non-convergence. There are other forms of higher order dependence assumptions that might also be necessary for a particular data set. However, the new specifications have proven very adequate. [?] shows that the models containing these new dependence parameters perform dramatically better than Markov models in terms of convergence, when applied to a number of classic small-scale network data sets. [?] fits the new specifications to a network of over 1000 nodes and shows how to assess model fit across many graph features.

5 Simulation, Estimation, and Assessment

It is relatively straightforward to simulate p^* models, and estimate parameters (as mentioned below), using long-established statistical approaches such the Metropolis algorithm [20] implementation of a Markov chain Monte Carlo. As first noted by [10], if the model is not degenerate, the algorithm will "burn-in" to a stationary distribution of graphs reflecting the parameter values in the model. The length of the burn-in depends on the starting graph for the simulation, the complexity of the model, and the size of the network. For small networks of 30 nodes, for instance, non-degenerate models can burnin within a few tens of thousands of iterations, which can be achieved within seconds on a fast enough computer. It is then possible to sample a number of graphs from this distribution and look at typical features of them, for instance, the density, the geodesic distribution, the frequencies of various triads, and so on [13]. In other words, although the model is based on certain configurations, the graphs from the distribution typically will exhibit certain other features of interest that can be investigated.

These models are especially appealing not only because they are readily simulated but also because the parameters can be estimated from available data. In the past, p^* models were fitted using pseudo-likelihood estimation based on logistic regression procedures ([6]; see [10] for a review). Although pseudo-likelihood can provide information about the data, especially in terms of identifying major effects [19], when models are close to degeneracy or when dependency is strong, the precise pseudo-likelihood parameter estimates are likely to be misleading.

A more reliable way to fit the models is through Markov chain Monte Carlo Maximum Likelihood Estimation (MCMCMLE). There are various algorithms possible to do this (see [22, 20]). While the technical details are complicated, the underlying conceptual basis is straightforward. MCMCMLE is based on simulation (hence, the MCMC part of the acronym). A distribution of graphs is simulated from an initial guess at parameter estimates. A sample from the resulting graph distribution is compared to the observed graph to see how well the observed graph is reproduced by the modeled configurations. If it is not well-reproduced the parameter estimates are appropriately adjusted. If the model is well-behaved, this procedure usually results in increasingly refined parameter estimates, until finally the procedure stops under some criterion. We do note one large differences between Markov models and models containing parameters from the new specifications: the new specifications are more likely to be well-behaved and result in convergent parameter estimates.

Once estimates have been obtained, the model can be simulated and assessed. The assessment is accomplished by comparing a statistic calculated from the observed graph to the distribution of the statistic generated by the model. This can be seen as a (rather demanding) goodness of fit diagnostic for the model. [?] shows how this approach can be used to improve models by the addition of extra effects. It is also an approach that permits judgments about how well competing models might represent the network.

We note that currently, there are three programs publicly available for the simulation, estimation, and goodness of fit of p^* models:

- the *StOCNET* suite of programs from the University of Groningen
  http://stat.gamma.rug.nl/stocnet/ (especially *SIENA*)
- the *statnet* program from the University of Washington
  http://csde.washington.edu/statnet
- the *pnet* program from the University of Melbourne
  http://www.sna.unimelb.edu.au

6   New Ideas — Network Imputation

One of the most interesting developments in the statistical modeling of networks centers on an approach to estimate missing nodes and missing links. We refer to it as *network imputation*.

To outline these ideas, first suppose a given network $\mathcal{X}$ is fit with a particular statistical model containing a set of network parameters collected in the vector θ. The likelihood given the data can then be denoted as

$$L(\theta|\mathbf{X} = \mathbf{x}),$$

where the dependence of the likelihood on the particular model is suppressed. The parameters in θ are estimated as discussed earlier, with the associated estimates denoted by $\hat{\theta}$. The likelihood evaluated at the estimated parameters is

$$L(\mathbf{x}; \hat{\boldsymbol{\theta}}).$$

In general, for a specific network, a model is adopted and a likelihood calculated. The likelihood is clearly dependent on the set of parameters that are chosen in the model. And if the observed network changes, so does the likelihood function. For our purposes, we will view the likelihood as a function not only of the parameters, but also of the data $\mathbf{x}$.

One of the most difficult problems in network analysis is determining whether the modeled network $\mathbf{X}$ contains the complete set of nodes, $\mathcal{N}_c$, and the complete set of edges, $\mathcal{E}_c$. So, let us assume that the complete node set is

$$\mathcal{N}_c = \mathcal{N}_o \cup \mathcal{N}_m$$

and the complete edge set is

$$\mathcal{E}_c = \mathcal{E}_o \cup \mathcal{E}_m$$

where $\mathcal{N}_o$ and $\mathcal{E}_o$ are the observed set of nodes and edges, respectively, while $\mathcal{N}_m$ and $\mathcal{E}_m$ are the missing (that is, unobserved) set of nodes and edges, respectively. If $\mathcal{N}_m = \emptyset$ ($\mathcal{E}_m = \emptyset$), then all nodes (edges) are observed in $\mathbf{x}$; otherwise, some nodes (edges) are missing and the goal is to impute the missing nodes and/or edges. For applications, it is of interest to estimate the components of $\mathcal{N}_m$ and $\mathcal{E}_m$.

We briefly describe one technique for estimating missing edges for a fixed set of nodes.

Estimating missing edges

We briefly outline the procedure we have used to estimate missing links. We note that the distribution of $\mathbf{X}$ is based on an approximate multivariate normality assumption of particular graph statistics. Specifically, we select a number of graph statistics, which, as we describe in Steinley and Wasserman (2006), are approximately Gaussian after transformations. We use only these statistics and assume they are statistically independent (which of course is probably far from the truth, but we view this work as just a first attempt at this). We then get a joint probability function and from that, a function that we can maximize (with respect to one missing link at a time). We assume a link is missing, and we add the link that maximizes the function.

We of course are assuming that the statistics chosen are the sufficient statistics for some underlying probability distribution.

The details follow:

1. Posit the distribribution, $\mathcal{D}$, of $\mathbf{X}$
2. Based on $\mathcal{D}$, compute the initial likelihood of $\mathbf{X} = \mathbf{x}$ using the current estimates of θ, $L(\mathbf{x}; \hat{\boldsymbol{\theta}})$, denoted as L.
3. Set $j = 1$. Let $L_j = L$.
4. The total possible number of edges in $\mathcal{E}_c$ is $\binom{g}{2}$. Let the number of observed edges be O, consequently resulting in the number of unobserved edges as $\binom{g}{2} - O$.

5. For the ith $\left(i = 1, \ldots, \binom{g}{2} - O\right)$ unobserved edge in $\mathcal{E}_m$ (denoted as $\mathbf{X}^{(i)}$, change $\mathbf{X}^{(i)} = 0$ to $\mathbf{X}^{(i)} = 1$, reestimate θ (denoted as $\hat{\boldsymbol{\theta}}^{(i)}$), and compute the associated likelihood $L_j^{(i)}$.
6. Repeat step 3 for all $i \in \mathcal{E}_m$.
7. Choose $L_j = \max_i L_j^{(i)}$
8. If $L_j > L$, permanently change $\mathbf{X}^{(i)}$ from 0 to 1; set $j = j + 1$ and $L = L_j$; repeat steps 3–7.
9. If $L_j \leq L$ stop adding edges to $\mathbf{X}$.

The basic idea is to incrementally "test" each edge that has not been observed to see if the addition of the link to the edge set increases the likelihood of the observed network. Additions of edges continues until the likelihood of the network can no longer be increased or until the incremental increase is small.

The procedure described here is the simplest way to look for missing edges within a network; however, there are many modifications that can made to search for links. One of the most obvious (and perhaps most worth considering) is the possibility of adding multiple links at the same time and evaluate the resulting likelihood. For instance, the likelihood could also be evaluated when all possible pairs of links are also added to the graph. In fact, it would be possible to add triplets, quadruplets, etc., up to the logical conclusion of all missing nodes being added simultaneously and computing the likelihood of the complete graph. This procedure would be a complete enumeration task, becoming combinatorically infeasible for networks with any reasonably sized node set.

In addition to adding links, it would also be possible to add nodes to the network. The proposed algorithm for adding links to a network would become augmented where a sequential addition of nodes is simultaneously evaluated. Thus, one could consider the likelihood of the network with g nodes and then consider the likelihood of a network with $g + 1$ nodes and its possible links. Prior to conducting this procedure, the researcher is required to determine the maximum possible number of nodes that will be added to the observed network. Finally, it should be recognized that the described procedures for detecting missing nodes and links can be augmented *ad infinitum* to adapt to specific network structures hypothesized by researchers. The only caveat is that additional structural properties imposed by the researcher can have effects (possibly adverse) on the final solution.

Acknowledgements

This research was supported by grants from the National Science Foundation and the US Office of Naval Research, and the Australian Research Council. We thank Stephen Fienberg for comments on this chapter.

References

[1] Wasserman, S., and Faust, K. (1994). *Social Network Analysis: Methods and Applications*. New York: Cambridge University Press.

[2] Carrington, P.J., Scott, J., and Wasserman, S. (eds.) (2005). *Models and Methods in Social Network Analysis*. New York: Cambridge University Press

[3] Snijders, T.A.B., Steglich, C., & Schweinberger, M. (in press). Modeling the co-evolution of networks and behavior. To appear in van Monfort et al (Eds.), *Longitudinal Models in the Behavioral and Related Sciences*. New York: Erlbaum.

[4] Hoff, P., Raftery, A., & Handcock, M. (2002). Latent space approaches to social network analysis. *Journal of the American Statistical Association, 97*, 1090-1098

[5] Frank, O., and Strauss, D. (1986). Markov graphs. *Journal of the American Statistical Association. 81*, 832–842.

[6] Strauss, D., & Ikeda, (1990). Pseudo-likelihood estimation for social networks. *Journal of the American Statistical Association. 85*, 204-212.

[7] Wasserman, S., and Pattison, P.E. (1996). Logit models and logistic regressions for social networks: I. An introduction to Markov random graphs and p^*. *Psychometrika. 60*, 401–426.

[8] Pattison, P.E., and Wasserman, S. (1999). Logit models and logistic regressions for social networks: II. Multivariate relations. *British Journal of Mathematical and Statistical Psychology. 52*, 169–193.

[9] Robins, G.L., Pattison, P.E., and Wasserman, S. (1999). Logit models and logistic regressions for social networks, III. Valued relations. *Psychometrika. 64*, 371–394

[10] Anderson, C.J., Wasserman, S., and Crouch, B. (1999). A p^* primer: Logit models for social networks. *Social Networks. 21*, 37–66.

[11] Borner, K., Sanyal, S., & Vespignani, A. (in press). Network science: A theoretical and practical framework. In Blaise Cronin (Ed.), *Annual Review of Information Science & Technology*, Volume 4. Medford, NJ: Information Today, Inc./American Society for Information Science and Technology.

[12] Wasserman, S., and Robins, G.L. (2005). An introduction to random graphs, dependence graphs, and p^*. In Carrington, P.J., Scott, J., and Wasserman, S. (eds.), *Models and Methods in Social Network Analysis*. New York: Cambridge University Press.

[13] Robins, G.L., Pattison, P.E., Kalish, Y., & Lusher, D. (in press). An introduction to exponential random graph (p^*) models for social networks. *Social Networks.*

[14] Hunter, D.R. (In press). Curved exponential family models for social networks. *Social Networks*

[15] Robins, G.L., Elliot, P., and Pattison, P.E. (2001). Network models for social selection processes. *Social Networks. 23*, 1–30 and *Psychometrika. 66*, 161–190.

[16] Handcock, M.S. (2002). Statistical models for social networks: Degeneracy and inference. In Breiger, R., Carley, K., & Pattison, P. (eds.). *Dynamic Social Network Modeling and Analysis* (pp. 229-240). Washington DC: National Academies Press.

[17] Park, J., & Newman, M. (2004). Solution of the 2-star model of a network. *Physical Review E, 70*, 066146.

[18] Robins, G.L., Pattison, P.E., & Woolcock, J. (2005). Social networks and small worlds. *American Journal of Sociology. 110*, 894-936.

[19] Robins, G.L., Snijders, T.A.B., Wang, P., Handcock, M., & Pattison, P.E. (In press). Recent developments in exponential random graph (p^*) models for social networks. *Social Networks.*

[20] Snijders, T.A.B. (2002). Markov chain Monte Carlo estimation of exponential random graph models. *Journal of Social Structure. 3*, 2.

[21] Snijders, T.A.B., Pattison, P.E., Robins, G.L., & Handcock, M. (2006). New specifications for exponential random graph models. *Sociological Methodology.*

[22] Hunter, D. & Handcock, M. (2006). Inference in curved exponential family models for networks. *Journal of Computational and Graphical Statistics. 15*, 565–583.

[23] Holland, P.W., and Leinhardt, S. (1977). Notes on the statistical analysis of social network data.

[24] Holland, P. W., and Leinhardt, S. (1981). An exponential family of probability distributions for directed graphs. *Journal of the American Statistical Association. 76*, 33–65 (with discussion).

Combining Stochastic Block Models and Mixed Membership for Statistical Network Analysis

Edoardo M. Airoldi[1,*], David M. Blei[2], Stephen E. Fienberg[1,3],
and Eric P. Xing[1]

[1] School of Computer Science, Carnegie Mellon University,
Pittsburgh PA 15213 USA
[2] Department of Computer Science, Princeton University,
Princeton NJ 08540 USA
[3] Department of Statistics, Carnegie Mellon University,
Pittsburgh PA 15213 USA
eairoldi@cs.cmu.edu

Abstract. Data in the form of multiple matrices of relations among objects of a single type, representable as a collection of unipartite graphs, arise in a variety of biological settings, with collections of author-recipient email, and in social networks. Clustering the objects of study or situating them in a low dimensional space (e.g., a simplex) is only one of the goals of the analysis of such data; being able to estimate relational structures among the clusters themselves may be important. In [1], we introduced the family of *stochastic block models of mixed membership* to support such integrated data analyses. Our models combine features of mixed-membership models and block models for relational data in a hierarchical Bayesian framework. Here we present a *nested* variational inference scheme for this class of models, which is necessary to successfully perform fast approximate posterior inference, and we use the models and the estimation scheme to examine two data sets. (1) a collection of sociometric relations among monks is used to investigate the crisis that took place in a monastery [2], and (2) data from a school-based longitudinal study of the health-related behaviors of adolescents. Both data sets have recently been reanalyzed in [3] using a latent position clustering model and we compare our analyses with those presented there.

1 Introduction

Relational information arise in a variety of settings, e.g., in scientific literature papers are connected by citation, in the word wide web the webpages are connected by hyperlinks, and in cellular systems the proteins are often related by physical protein-protein interactions revealed in yeast-two-hybrid experiments. These types of relational data violate the assumptions of independence or exchangeability of objects adopted in many conventional analyses. In fact, the relationships themselves between objects are often of interest in addition to the

---

* To whom correspondence should be addressed, edo@cmu.edu.

E.M. Airoldi et al. (Eds.): ICML 2006 Ws, LNCS 4503, pp. 57–74, 2007.
© Springer-Verlag Berlin Heidelberg 2007

object attributes. For example, one may be interested in predicting the citations of newly written papers or the likely links of a web-page, or in clustering cellular proteins based on patterns of interactions between them.

In many such applications, clustering the objects of study or projecting them in a low dimensional space (e.g., a simplex) is only one of the goals of the analysis. Being able to estimate the relational structures among the clusters themselves is often as important as object clustering. For example, from observations about email communications of a study population, one may be not only interested in identifying groups of people of common characteristics or social states, but also at the same time exploring how the overall communication volume or pattern among these groups can reveal the organizational structures of the population. A popular class of probabilistic models for relational data analysis are based on the stochastic block model (SBM) formalism for psychometric and sociological analysis pioneered by Holland and Leinhardt [4], and later extended in various contexts [5,6,7,8,9]. In machine learning, Markov random networks have been used for link prediction [10] and the traditional block models have been extended to include nonparametric Bayesian priors [11,12] and to integrate relations and text [13]. Typically, these models posit that every node in a study network is characterized by a unary *latent aspect* that accounts for its interaction patterns to peers in the networks; and conditioning on the observed network topology one can reason about these *latent aspects* of nodes via posterior inference.

Largely disjoint from the network analysis literature, methodologies for latent aspect modeling have also been widely investigated in the contexts of different informational retrieval problems concerning modeling the high-dimensional non-relational attributes such as text content or genetic-allele profile. In many of these domains, variants of a mixed membership formalism have been proposed to capture a more realistic assumption about the observed attributes, that the observations are resulted from contributions from multiple latent aspects rather than a unary aspects as assumed in most extant network models such as SBM. The mixed membership models have emerged as a powerful and popular analytical tool for analyzing large databases involving text [14], text and references [15,16], text and images [17], multiple disability measures [18,19], and genetics information [20,21,22]. These models often employ a simple generative model, such as a bag-of-words model or a naive Bayes, embedded in a hierarchical Bayesian framework involving a latent variable structure that combines multiples latents aspects. This scheme induces dependencies among the objects' relational behaviors in the form of probabilistic constraints over the estimation of what might otherwise be an extremely large set of parameters.

In modern network analysis tasks described above, it is desirable to also relax the unary-aspect assumption on each node imposed by extant models. We have proposed a new class of stochastic network models based the principle of *stochastic block models of mixed membership* [1], which combines features of the mixed-membership models [18] and the block models [23,24,25,9] via a hierarchical Bayesian framework, and offers a flexible machinery to capture rich semantic aspects of various network data. In this paper, we describe an instantiation of this class of model, referred to as *admixture of latent blocks* (ALB) [26] to reasons to be explained shortly, for analyzing networks of objects with multiple latent roles (e.g., social activities in case the objects refer to people, or biological functions in case the objects refer to proteins). As mentioned above, classical network models such as the stochastic block models only allow each nodes to bear a single role. Our model alleviates this constraint, and furthermore posits that each nodes can adopt different roles when interacting with different other nodes.

Here is an outline of the rest of the paper. In Sections 2 we present the statistical formulation of the *Admixture of Latent Blocks* model (ALB). Then, in Section 3, we describe a variational inference algorithm for latent role inference and approximate maximum likelihood parameter estimation. In Section 4, we apply our model to two social networks widely studied in the literature, and we compare results of our analysis with that from a latent space model recently developed by Handcock, et al. [3].

2   The Statistical Model

We concern ourselves with modeling data represented as a collection of directed unipartite graphs. A unipartite graph is a graph whose nodes are of a single type, e.g., individual human beings in case of a person-to-person communication network, as opposed to bipartite and multipartite graphs, where the nodes are of two or multiple types (e.g., genes-to-experiments [14,27] or employees-to-tasks-to-resources [28]).

Let $G = (N, R)$ denote a graph with edge set R on node set N. We consider situations where we observe a collection of M unipartite graphs, $\mathcal{G} = \{G_m : r = 1, \ldots, M\}$ defined on a common set of nodes $\mathcal{N}$, of which the presence or absence of edges between node-pair i and j in graph G_m is denoted by variable $R_m(p, q)$. For example, in our experiment presented in the sequel, $\mathcal{N}$ corresponds a group of monks in a monestary [2], and $\{R_m(p, q)\}$ correspond to the relationships measured among these monks over a period. We observe typically asymmetric binary relations such as "Do you like X?", over a sequence of time.

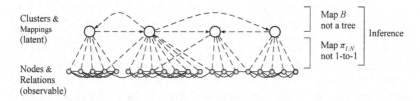

Fig. 1. The scientific problem at a glance. The goal of the analysis is to make inference on two mappings; nodes-to-clusters (via $\pi_{1:N}$) and clusters-to-clusters (via B). The facts that B does not necessarily encode a tree, and that $\pi_{1:N}$ is not necessarily one-to-one distinguish our formulation from typical hierarchical and hard clustering.

The analysis of such data typically focuses on the following objectives: (1) identifying clustering of nodes; (2) determining the number of clusters; and (3) estimating the probability distribution of interactions among actors within and between clusters. Back to the example of the monestary social network, objective 1 translates to identifying the solid factions among monks, In addition one wants to determine how many factions are likely to exist in the monastery, and how the factions relate to one another.

2.1   The Model

Our approach detailed below employs a hierarchical Bayesian formalism that encodes statistical assumptions underlying a network generative process. This process generates the observed networks according to the latent distribution of the hypothetical group-involvement of each monk, as specified by a mixed-memembership multinomial vector $\pi := [\pi_1, \ldots, \pi_K]'$ where π_i denote the probability of a monk belonging to group i; and the probabilities of having interactions between different groups, as defined by a matrix of Bernoulli rates $B_{(K \times K)} = \{B_{ij}\}$ where B_{ij} represents the probability of having a link between a monk from group i and a monk from group j. Each monk is associated with a unique π, meaning that he can be simultaneously belonging to multiple groups, and the degree of involvements in different groups is unique for each monk; and π of different monks independently follow a Dirichlet distribution parameterized by α.

More generally, for graph m and each node, let indicator vector [1] $z_{p \to q}^m$ denote the group membership of node p when it is to approach with node q; let $z_{p \leftarrow q}^m$ denote the group membership of node q when it is approached by node p; let $N := |\mathcal{N}|$ denote the number of nodes in the graph; and let K denote the number of distinct groups a node can belong to. An admixture of latent blocks (ALB) model posit that a sequence of M networks can be instantiated according to the following procedure:

- For each node $p = 1, \ldots, N$:
    - $\pi_p \sim$ Dirichlet (α) sample a K dimensional *mixed membership* vector;
- for each network G_m, and each pair of nodes $(p, q) \in [1, N] \times [1, N]$ (denote p as the *initiator* and q as the *receiver*) in G_m:
    - $z_{p \to q}^m \sim$ Multinomial (π_p) sample membership indicator for the initiator,
    - $z_{p \leftarrow q}^m \sim$ Multinomial (π_q) sample membership indicator for the receiver,
    - $R_m(p, q) \sim$ Bernoulli $(z_{p \to q}^{m\top} B\, z_{p \leftarrow q}^m)$ sample the value of their interaction.

It is noteworthy that in the above model, the group membership of each node is *context dependent*, that is, each nodes can assume different membership when interacting to or being interacted by different peers. Therefore, each node is statistically an admixture of group-specific interactions, and we denote the two sets of latent group indicators corresponding to the *m-th* observed network by

---

[1] An indicator vector of memberships in one of the K groups is defined as a K-dimensional vector of which only one element whose index corresponds to the id of the group to be indicated equals to one, and all other elements equal to zero.

$\{z_{p \rightarrow q}^m : p, q \in \mathcal{N}\} =: Z_m^{\rightarrow}$ and $\{z_{p \leftarrow q}^m : p, q \in \mathcal{N}\} =: Z_m^{\leftarrow}$. Marginalizing out the latent group indicators, it is easy to show that the probability of observing an interaction between node p and q across the M networks is $\bar{\sigma}_{pq} = \boldsymbol{\pi}_p^{\top} B \boldsymbol{\pi}_q$.

Under an ALB model outlined above, the joint probability distribution of the data, $R_{1:M}$, and the latent variables $(\boldsymbol{\pi}_{1:N}, Z_{1:M}^{\rightarrow}, Z_{1:M}^{\leftarrow})$ can be written in the following factored form:

$$p(R_{1:M}, \boldsymbol{\pi}_{1:N}, Z_{1:M}^{\rightarrow}, Z_{1:M}^{\leftarrow} | \alpha, B) \tag{1}$$
$$= \prod_{m} \prod_{p,q} P(R_m(p,q)|z_{p \rightarrow q}^m, z_{p \leftarrow q}^m, B) P(z_{p \rightarrow q}^m|\boldsymbol{\pi}_p) P(z_{p \leftarrow q}^m|\boldsymbol{\pi}_q) \prod_{p} p_3(\boldsymbol{\pi}_p|\alpha).$$

To compute the likelihood of the observed networks, one needs to marginalize out the hidden variables $\boldsymbol{\pi}$ and Z for all notes, which is intractable for even for small graphs. In §3, we describe a variational scheme to approximate this likelihood for parameter estimation.

2.2   Dealing with Sparsity

Most networks in real world are sparse, meaning that most pairs of nodes do not have edges connecting them. But in many network analyses, observations about interactions and non-interactions are equally important in terms of their contributions to model fitness. In other words, they would compete for a statistical explanation in terms of estimates for parameters (α, B), and would both influence the distribution of latent variables such as $\boldsymbol{\pi}_{1:N}$. A non desirable consequence of this, in scenarios where interactions are rare, is that parameter estimation and posterior inference would explain patterns of non-interaction rather than patterns of interaction.

In order to be able to calibrate the importance of rare interactions, we introduce the sparsity parameter $\rho \in [0, 1]$, which models how often a non-interaction is due to measurement noise (which is common in certain experimentally derived networks such as the protein-protein interaction networks) and how often it carries information about the group memberships of the nodes. This leads to a small extension of the generative process outlined in the last subsection. Specifically, instead of drawing an edge directly from a Bernoulli with rate $z_{p \rightarrow q}^{m\ \top} B\ z_{p \leftarrow q}^m$, now we sample an interaction with probability $\sigma_{pq}^m = (1 - \rho) \cdot z_{p \rightarrow q}^{m\ \top} B\ z_{p \leftarrow q}^m$; therefore the probability of having no interaction this pair of nodes is $1 - \sigma_{pq}^m = (1 - \rho) \cdot z_{p \rightarrow q}^{m\ \top} (1 - B)\ z_{p \leftarrow q}^m + \rho$. This is equivalent to re-parameterizing the interaction matrix B. During estimation and inference, a large value of ρ would cause the interactions in the matrix to be weighted more than non-interactions in determining the estimates of $(\alpha, B, \boldsymbol{\pi}_{1:N})$.

3   Parameter Estimation and Posterior Inference

We use an empirical Bayes framework for estimating the parameters (α, B), and employ a mean-field approximation scheme [29] for posterior inference of the (latent) mixed-membership vectors, $\pi_{1:N}$. Model selection can be performed to determine the plausible value of K—the number of groups of nodes—based on a strategy described in [30].

In order to estimate (α, B) and infer the posterior distributions of $\pi_{1:N}$ we need to be able to evaluate the likelihood, which involves the non-tractable integral over Z and $\pi_{1:N}$ in Equation 1. Given the large amount of data available for most networks, we focus on approximate posterior inference strategies in the context of variational methods, and we find a tractable lower bound for the likelihood that can be used as a surrogate for inference purposes. This leads to approximate MLEs for the hyper-parameters and approximate posterior distributions for the (latent) mixed-membership vectors.

3.1   Lower Bound for the Likelihood

According to the mean-field theory [29,31], one can approximate an intractable distribution such as the one defined by Equation (1) by a fully factored distribution $q(\pi_{1:N}, Z_{1:M}^{\rightarrow}, Z_{1:M}^{\leftarrow})$ defined as follows:

$$q(\pi_{1:N}, Z_{1:M}^{\rightarrow}, Z_{1:M}^{\leftarrow} | \gamma_{1:N}, \Phi_{1:M}^{\rightarrow}, \Phi_{1:M}^{\leftarrow})$$
$$= \prod_p q_1(\pi_p | \gamma_p) \prod_m \prod_{p,q} \left(q_2(z_{p\rightarrow q}^m | \phi_{p\rightarrow q}^m, 1) \, q_2(z_{p\leftarrow q}^m | \phi_{p\leftarrow q}^m, 1) \right), \quad (2)$$

where q_1 is a Dirichlet, q_2 is a multinomial, and $\Delta = (\gamma_{1:N}, \Phi_{1:M}^{\rightarrow}, \Phi_{1:M}^{\leftarrow})$ represent the set of free *variational parameters* need to be estimated in the approximate distribution.

Minimizing the Kulback-Leibler divergence between this $q(\pi_{1:N}, Z_{1:M}^{\rightarrow}, Z_{1:M}^{\leftarrow} | \Delta)$ and the original $p(\pi_{1:N}, Z_{1:M}^{\rightarrow}, Z_{1:M}^{\leftarrow})$ defined by Equation (1) leads to the following approximate lower bound for the likelihood.

$$\mathcal{L}_\Delta(q, \Theta) = \mathbb{E}_q \left[\log \prod_m \prod_{p,q} p_1(R_m(p,q) | z_{p\rightarrow q}^m, z_{p\leftarrow q}^m, B) \right]$$
$$+ \mathbb{E}_q \left[\log \prod_m \prod_{p,q} p_2(z_{p\rightarrow q}^m | \pi_p, 1) \right] + \mathbb{E}_q \left[\log \prod_m \prod_{p,q} p_2(z_{p\leftarrow q}^m | \pi_q, 1) \right]$$
$$+ \mathbb{E}_q \left[\log \prod_p p_3(\pi_p | \alpha) \right] - \mathbb{E}_q \left[\prod_p q_1(\pi_p | \gamma_p) \right]$$
$$- \mathbb{E}_q \left[\log \prod_m \prod_{p,q} q_2(z_{p\rightarrow q}^m | \phi_{p\rightarrow q}^m, 1) \right] - \mathbb{E}_q \left[\log \prod_m \prod_{p,q} q_2(z_{p\leftarrow q}^m | \phi_{p\leftarrow q}^m, 1) \right].$$

Working on the single expectations leads to the following expression,

$$\mathcal{L}_\Delta(q, \Theta) = \sum_m \sum_{p,q} \sum_{g,h} \phi_{p\rightarrow q,g}^m \phi_{p\leftarrow q,h}^m \cdot f\left(R_m(p,q), B(g,h) \right)$$
$$+ \sum_m \sum_{p,q} \sum_g \phi_{p\rightarrow q,g}^m \left[\psi(\gamma_{p,g}) - \psi(\sum_g \gamma_{p,g}) \right]$$
$$+ \sum_m \sum_{p,q} \sum_h \phi_{p\leftarrow q,h}^m \left[\psi(\gamma_{p,h}) - \psi(\sum_h \gamma_{p,h}) \right]$$

$$+ \sum_p \log \Gamma(\sum_k \alpha_k) - \sum_{p,k} \log \Gamma(\alpha_k) + \sum_{p,k} (\alpha_k - 1) \left[\psi(\gamma_{p,k}) - \psi(\sum_k \gamma_{p,k}) \right]$$

$$- \sum_p \log \Gamma(\sum_k \gamma_{p,k}) + \sum_{p,k} \log \Gamma(\gamma_{p,k}) - \sum_{p,k} (\gamma_{p,k} - 1) \left[\psi(\gamma_{p,k}) - \psi(\sum_k \gamma_{p,k}) \right]$$

$$- \sum_m \sum_{p,q} \sum_g \phi^m_{p \to q,g} \log \phi^m_{p \to q,g} - \sum_m \sum_{p,q} \sum_h \phi^m_{p \leftarrow q,h} \log \phi^m_{p \leftarrow q,h}$$

where

$$f\left(R_m(p,q), B(g,h) \right) = R_m(p,q) \log B(g,h) + \left(1 - R_m(p,q) \right) \log \left(1 - B(g,h) \right);$$

m runs over $1, \dots, M$; p, q run over $1, \dots, N$; g, h, k run over $1, \dots, K$; and $\psi(x)$ is the derivative of the log-gamma function, $\frac{d \log \Gamma(x)}{dx}$.

3.2 The Expected Value of the Log of a Dirichlet Random Vector

The computation of the lower bound for the likelihood requires us to evaluate $\mathbb{E}_q \left[\log \pi_p \right]$ for $p = 1, \dots, N$. Recall that the density of an exponential family distribution with natural parameter θ can be written as

$$p(x|\alpha) = h(x) \cdot c(\alpha) \cdot \exp \left\{ \sum_k \theta_k(\alpha) \cdot t_k(x) \right\}$$

$$= h(x) \cdot \exp \left\{ \sum_k \theta_k(\alpha) \cdot t_k(x) - \log c(\alpha) \right\}.$$

Omitting the node index p for convenience, we can rewrite the density of the Dirichlet distribution p_3 as an exponential family distribution,

$$p_3(\pi|\alpha) = \exp \left\{ \sum_k (\alpha_k - 1) \log(\pi_k) - \log \frac{\prod_k \Gamma(\alpha_k)}{\Gamma(\sum_k \alpha_k)} \right\},$$

with natural parameters $\theta_k(\alpha) = (\alpha_k - 1)$ and natural sufficient statistics $t_k(\pi) = \log(\pi_k)$. Let $c'(\theta) = c(\alpha_1(\theta), \dots, \alpha_K(\theta))$; using a well known property of the exponential family distributions [32] we find that

$$\mathbb{E}_q \left[\log \pi_k \right] = \mathbb{E}_\theta \left[\log t_k(x) \right] = \psi\left(\alpha_k \right) - \psi\left(\sum_k \alpha_k \right),$$

where $\psi(x)$ is the derivative of the log-gamma function, $\frac{d \log \Gamma(x)}{dx}$.

3.3 Variational E Step

The approximate lower bound for the likelihood $\mathcal{L}_\Delta(q, \Theta)$ can be maximized using exponential family arguments and coordinate ascent [33].

Isolating terms containing $\phi^m_{p \to q,g}$ and $\phi^m_{p \leftarrow q,h}$ we obtain $\mathcal{L}_{\phi^m_{p \to q,g}}(q, \Theta)$ and $\mathcal{L}_{\phi^m_{p \to q,g}}(q, \Theta)$. The natural parameters $g^m_{p \to q}$ and $g^m_{p \leftarrow q}$ corresponding to the natural sufficient statistics $\log(z^m_{p \to q})$ and $\log(z^m_{p \leftarrow q})$ are functions of the other latent variables and the observations. We find that

$$g^m_{p \to q,g} = \log \pi_{p,g} + \sum_h z^m_{p \leftarrow q,h} \cdot f\left(R_m(p,q), B(g,h)\right),$$

$$g^m_{p \leftarrow q,h} = \log \pi_{q,h} + \sum_g z^m_{p \to q,g} \cdot f\left(R_m(p,q), B(g,h)\right),$$

for all pairs of nodes (p,q) in the m-th network; where $g, h = 1, \ldots, K$, and

$$f\left(R_m(p,q), B(g,h)\right) = R_m(p,q) \log B(g,h) + \left(1 - R_m(p,q)\right) \log\left(1 - B(g,h)\right).$$

This leads to the following updates for the variational parameters $(\phi^m_{p \to q}, \phi^m_{p \leftarrow q})$, for a pair of nodes (p,q) in the m-th network:

$$\hat\phi^m_{p \to q,g} \propto e^{\mathbb{E}_q \left[g^m_{p \to q,g}\right]} \tag{3}$$

$$= e^{\mathbb{E}_q \left[\log \pi_{p,g}\right]} \cdot e^{\sum_h \phi^m_{p \leftarrow q,h} \cdot \mathbb{E}_q \left[f\left(R_m(p,q), B(g,h)\right)\right]}$$

$$= e^{\mathbb{E}_q \left[\log \pi_{p,g}\right]} \cdot \prod_h \left(B(g,h)^{R_m(p,q)} \cdot \left(1 - B(g,h)\right)^{1 - R_m(p,q)}\right)^{\phi^m_{p \leftarrow q,h}}$$

$$\hat\phi^m_{p \leftarrow q,h} \propto e^{\mathbb{E}_q \left[g^m_{p \leftarrow q,h}\right]} \tag{4}$$

$$= e^{\mathbb{E}_q \left[\log \pi_{q,h}\right]} \cdot e^{\sum_g \phi^m_{p \to q,g} \cdot \mathbb{E}_q \left[f\left(R_m(p,q), B(g,h)\right)\right]}$$

$$= e^{\mathbb{E}_q \left[\log \pi_{q,h}\right]} \cdot \prod_g \left(B(g,h)^{R_m(p,q)} \cdot \left(1 - B(g,h)\right)^{1 - R_m(p,q)}\right)^{\phi^m_{p \to q,g}}$$

for $g, h = 1, \ldots, K$. These estimates of the parameters underlying the distribution of the nodes' group indicators $\phi^m_{p \to q}$ and $\phi^m_{p \leftarrow q}$ need be normalized, to make sure $\sum_k \phi^m_{p \to q,k} = \sum_k \phi^m_{p \leftarrow q,k} = 1$.

Isolating terms containing $\gamma_{p,k}$ we obtain $\mathcal{L}_{\gamma_{p,k}}(q, \Theta)$. Setting $\frac{\partial \mathcal{L}_{\gamma_{p,k}}}{\partial \gamma_{p,k}}$ equal to zero and solving for $\gamma_{p,k}$ yields:

$$\hat\gamma_{p,k} = \alpha_k + \sum_m \sum_q \phi^m_{p \to q,k} + \sum_m \sum_q \phi^m_{p \leftarrow q,k}, \tag{5}$$

for all nodes $p \in \mathcal{P}$ and $k = 1, \ldots, K$.

The t-th iteration of the variational E step is carried out for fixed values of $\Theta^{(t-1)} = (\alpha^{(t-1)}, B^{(t-1)})$, and finds the optimal approximate lower bound for the likelihood $\mathcal{L}_{\Delta^*}(q, \Theta^{(t-1)})$.

3.4   Variational M Step

The optimal lower bound $\mathcal{L}_{\Delta^*}(q^{(t-1)}, \Theta)$ provides a tractable surrogate for the likelihood at the t-th iteration of the variational M step. We derive empirical

Bayes estimates for the hyper-parameters Θ that are based upon it.[2] That is, we maximize $\mathcal{L}_{\Delta^*}(q^{(t-1)}, \Theta)$ with respect to Θ, given expected sufficient statistics computed using $\mathcal{L}_{\Delta^*}(q^{(t-1)}, \Theta^{(t-1)})$.

Isolating terms containing α we obtain $\mathcal{L}_\alpha(q, \Theta)$. Unfortunately, a closed form solution for the approximate maximum likelihood estimate of α does not exist [14]. We can produce a Newton-Raphson method that is linear in time, where the gradient and Hessian for the bound $\mathcal{L}_\alpha$ are

$$\frac{\partial \mathcal{L}_\alpha}{\partial \alpha_k} = N \left(\psi \left(\sum_k \alpha_k \right) - \psi(\alpha_k) \right) + \sum_p \left(\psi(\gamma_{p,k}) - \psi \left(\sum_k \gamma_{p,k} \right) \right),$$

$$\frac{\partial \mathcal{L}_\alpha}{\partial \alpha_{k_1} \alpha_{k_2}} = N \left(\mathbb{I}_{(k_1 = k_2)} \cdot \psi'(\alpha_{k_1}) - \psi' \left(\sum_k \alpha_k \right) \right).$$

Isolating terms containing B we obtain $\mathcal{L}_B$, whose approximate maximum is

$$\hat{B}(g, h) = \frac{1}{M} \sum_m \left(\frac{\sum_{p,q} R_m(p, q) \cdot \phi^m_{p \to qg} \, \phi^m_{p \leftarrow qh}}{\sum_{p,q} \phi^m_{p \to qg} \, \phi^m_{p \leftarrow qh}} \right), \tag{6}$$

for every index pair $(g, h) \in [1, K] \times [1, K]$.

In Section 2.2 we introduced an extra parameter, ρ, to control the relative importance of presence and absence of interactions in likelihood, i.e., the score that informs inference and estimation. Isolating terms containing ρ we obtain $\mathcal{L}_\rho$. We may then estimate the sparsity parameter ρ by

$$\hat{\rho} = \frac{1}{M} \sum_m \left(\frac{\sum_{p,q} \left(1 - R_m(p, q) \right) \cdot \left(\sum_{g,h} \phi^m_{p \to qg} \, \phi^m_{p \leftarrow qh} \right)}{\sum_{p,q} \sum_{g,h} \phi^m_{p \to qg} \, \phi^m_{p \leftarrow qh}} \right). \tag{7}$$

Alternatively, we can fix ρ prior to the analysis; the density of the interaction matrix is estimated with $\hat{d} = \sum_{m,p,q} R_m(p, q)/(N^2 M)$, and the sparsity parameter is set to $\tilde{\rho} = (1 - \hat{d})$. This latter estimator attributes all the information in the non-interactions to the point mass, i.e., to latent sources other than the block model B or the mixed membership vectors $\pi_{1:N}$. It does however provide a quick recipe to reduce the computational burden during exploratory analyses.[3]

3.5   Smoothing

In problems where the number of clusters is deemed to be likely large a-priori, we can smooth the (consequently large number of) cluster-to-cluster relation probabilities encoded in the block model B by positing that all the elements $B(g, h)$ of the block model are non-observable samples from a common (prior) distribution. In the admixture of latent blocks model we posit that $p(B|\lambda)$ is a collection non-symmetric beta distributions, with a pair of hyper-parameters λ common to all elements of B.

---

[2] We could term these estimates *pseudo* empirical Bayes estimates, since they maximize an approximate lower bound for the likelihood, $\mathcal{L}_{\Delta^*}$.

[3] Note that $\tilde{\rho} = \hat{\rho}$ in the case of single membership. In fact, that implies $\phi^m_{p \to qg} = \phi^m_{p \leftarrow qh} = 1$ for some (g, h) pair, for any (p, q) pair.

3.6   The Nested Variational EM Algorithm

The complete algorithm to perform variational inference in the model is described in detail in Figure 2. To achieve fast convergence, we employed a highly effective *nested* variational inference scheme based on a non-trivial scheduling of variational parameters updating. The resulting algorithm is also parallelizable on a computer cluster.

---

1. initialize $\gamma_{pk}^0 = \frac{2N}{K}$ for all p, k
2. **repeat**
3.   **for** $p = 1$ to N
4.   **for** $q = 1$ to N
5.   get **variational** $\phi_{p \to q}^{t+1}$ and $\phi_{p \leftarrow q}^{t+1} = f\left(R(p,q), \gamma_p^t, \gamma_q^t, B^t \right)$
6.   partially update $\gamma_p^{t+1}, \gamma_q^{t+1}$ and B^{t+1}
7. **until** convergence

---

1. initialize $\phi_{p \to q, g}^0 = \phi_{p \leftarrow q, h}^0 = \frac{1}{K}$ for all g, h
2. **repeat**
3.   **for** $g = 1$ to K
4.   update $\phi_{p \to q}^{s+1} \propto f_1\left(\phi_{p \leftarrow q}^s, \gamma_p, B \right)$
5.   normalize $\phi_{p \to q}^{s+1}$ to sum to 1
6.   **for** $h = 1$ to K
7.   update $\phi_{p \leftarrow q}^{s+1} \propto f_2\left(\phi_{p \to q}^s, \gamma_q, B \right)$
8.   normalize $\phi_{p \leftarrow q}^{s+1}$ to sum to 1
9. **until** convergence

Fig. 2. Top: The two-layered variational inference for $(\gamma, \phi_{p \to q, g}, \phi_{p \leftarrow q, h})$ and $M = 1$. The inner algorithm consists of Step 5. The function f is described in details in the bottom panel. The partial updates in Step 6 for γ and B refer to Equation 5 of Section 3.3 and Equation 6 of Section 3.4, respectively. **Bottom:** Inference for the variational parameters $(\phi_{p \to q}, \phi_{p \leftarrow q})$ corresponding to the basic observation $R(p, q)$. This nested algorithm details Step 5 in the top panel. The functions f_1 and f_2 are the updates for $\phi_{p \to q, g}$ and $\phi_{p \leftarrow q, h}$ described in Equations 3 and 4 of Section 3.3.

In a naïve iteration scheme for variational inference, one would initialize the variational Dirichlet parameters $\gamma_{1:N}$ and the variational multinomial parameters $(\phi_{p \to q}, \phi_{p \leftarrow q})$ to non-informative values, and then iterate until convergence the following two steps: (i) update $\phi_{p \to q}$ and $\phi_{p \leftarrow q}$ for all edges (p, q), and (ii) update γ_p for all nodes $p \in \mathcal{N}$. In such algorithm, at each variational inference cycle we need to allocate $NK + 2N^2K$ scalars. In our experiments [1] the naïve variational algorithm often failed to converge, or converged after a large number of iterations. We attribute this behavior to a dependence that our two main assumptions (block model and mixed membership) induce between $\gamma_{1:N}$ and B, which is not satisfied by the naïve algorithm. Some intuition about why this may happen follows. From a purely algorithmic perspective, the naïve variational EM algorithm instantiates a large coordinate ascent algorithm, where the parameters can be semantically divided into coherent blocks. Blocks are processed in a specific order, and the parameters within each block get all updated each time.[4] At every new iteration the naïve algorithm sets all the elements of $\gamma_{1:N}^{t+1}$ equal to the same constant. This dampens the likelihood by suddenly breaking the dependence between the estimates of parameters in $\widehat{\gamma}_{1:N}^{t}$ and in $\hat{B}^t$ that was being inferred from the data during the previous iteration.

Instead, the nested variational inference algorithm maintains some of this dependence that is being inferred from the data across the various iterations. This is achieved mainly through a different scheduling of the parameter updates in the various blocks. To a minor extent, the dependence is maintained by always keeping the block of free parameters, $(\phi_{p \to q}, \phi_{p \leftarrow q})$, optimized given the other variational parameters. Note that these parameters are involved in the updates of parameters in $\gamma_{1:N}$ and in B, thus providing us with a channel to maintain some of the dependence among them, i.e., by keeping them at their optimal value given the data. Further, the nested algorithm has the advantage that it trades time for space thus allowing us to deal with large graphs; at each variational cycle we need to allocate $NK + 2K$ scalars only. The increased running time is partially offset by the fact that the algorithm can be parallelized and leads to empirically observed faster convergence rates. This algorithm is also better than MCMC variations (i.e., blocked and collapsed Gibbs samplers) in terms of memory requirements and convergence rates.

4   Experiments: Applications to Social Networks

We illustrate our model and algorithm in the context of two examples that have recently been reanalyzed in [3] using a *latent position clustering model* and [34].

4.1   Example 1: Crisis in a Cloister

Sampson [2] surveyed 18 novice monks in a monastery and asked them to rank the other novices in terms of four *sociometric relations*: like/dislike, esteem, personal influence, and alignment with the monastic credo. Sampson's original analysis strongly suggested the existence of tight factions among the novices, and the events that took place during his stay at the monastery supported his observations. Briefly, novices of one faction left the monastery or were expelled over religious differences. The factions identified by Sampson provide a credible gold standard, to which we compare our results.

---

[4] Within a block, the order according to which (scalar) parameters get updated is not expected to affect convergence.

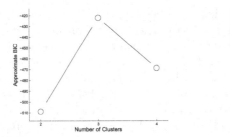

Fig. 3. The approximate BIC (left panel) suggests the relations among monks are best explained by a model with three factions. The faction-to-faction estimated relational patterns $\widehat{B}$ (right panel) suggest that the Outcasts are an isolated faction, whereas Young Turks *like* members of the Loyal Opposition, although the sentiment is not reciprocated.

We consider Breiger's collation of Sampson's data [35]. Briefly, for each of the four sociometric relations above, only the top three choices of each novice were recorded as positive relations—the edges in the graph. We use the following approximation to BIC for model selection:

$$BIC = 2 \cdot \log p(R) \approx 2 \cdot \log p(R|\widehat{\pi}, \widehat{Z}, \widehat{\alpha}, \widehat{B}) - |\alpha, B| \cdot \log |R|,$$

where $|\alpha, B|$ is the number of hyper-parameters in the model, and $|R|$ is the number of positive relations observed—following arguments in [3]. The approximate BIC value suggests that the relations among monks in the monastery studied by Sampson are best explained by a model with three factions, independently of the number of hyper-parameters in the ALB model we fit. Hence we fixed $\widehat{K} = 3$ in subsequent analyses, which involved ALB models with increasing degree of complexity. In the left panel of Figure 3 we show the approximate BIC for a model with a single hyper-parameter, α scalar. In the right panel of Figure 3 we show the estimated faction-to-faction block model, $\widehat{B}$, corresponds to a full model (i.e., no constraints on B). This estimate suggests that the Outcasts are an isolated faction, whereas Young Turks *like* members of the Loyal Opposition, although the sentiment is not reciprocated. In Figure 5 we investigate the the posterior means of the mixed membership scores, $\mathbb{E}[\pi|R]$, for the 18 monks in the monastery ($\alpha = 0.058$ scalar, $B := \mathbb{I}_3$). We have a panel for each monk, and the subscripts associated with the names of the monks specify the order according to which they left the monastery, e.g., John left first. The three factions on the X axis are the Outcast, the Young Turks , and the Loyal Opposition (from left to right); and on the Y axis we measure the degree of membership of monks to factions. From these panels, the centrality of the role played by John and Greg, first to leave the monastery, as well as the uncertain affiliations of Romul, and Victor to a minor extent, unequivocally emerge. The mixed membership vectors, $\pi_{1:18}$, provide us with low-dimensional representations of monks. In Figure 6 we plot them in their natural space, that

Fig. 4. Original matrix of sociometric relations (left), and estimated relations obtained by thresholding the posterior expectations $\pi_p{}'B\,\pi_q|R$ (center), and $\phi_p{}'B\,\phi_q|R$ (right)

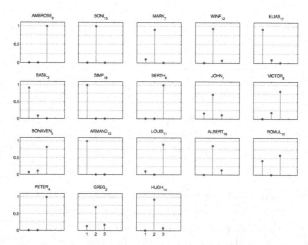

Fig. 5. The posterior mixed membership scores, π, for the 18 monks in the monastery. Each panel correspond to a monk; on the Y axis we measure the grade of membership, corresponding to the Outcast (left bar), to the Young Turks (center bar), and to the Loyal Opposition (right bar), on the X axis. The subscripts associated with the names of the monks specify the order according to which they left the monastery.

is, the(3-dimensional) simplex. Dots correspond to monks; the red circles were obtained by fixing $B = \mathbb{I}_3$ and $\alpha = 0.01$, whereas the blue triangles correspond to fixing $B := \mathbb{I}_3$, but estimating $\hat{\alpha} = 0.058$.

4.2   Example 2: Health-Related Behaviors of Adolescents

The National Longitudinal Study of Adolescent Health [36,37] includes questionnaire administered to a sample of students, who were allowed to nominate up to 10 friends. Following [3], we focus on friendship nominations collected among 71 students in grades 7 to 12 at one school. Two students did not nominate any friends, so we analyzed the network of (binary, asymmetric) friendship relations

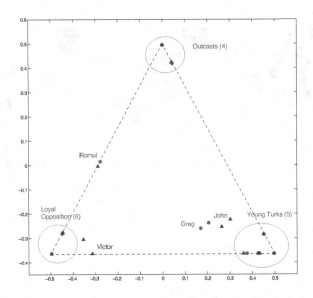

Fig. 6. Mixed membership vectors, $\pi_{1:18}$, plotted in the simplex. Points correspond to monks; the red circles correspond to an ALB model with ($B = \mathbb{I}_3, \alpha = 0.01$), whereas the blue triangles correspond to an ALB model with ($B := \mathbb{I}_3, \hat{\alpha} = 0.058$).

among the remaining 69 students. The left panel of Figure 8 shows the raw relations, and we contrast this to the estimated networks in the central and right panels based on our model estimates using the full model. We proceeded with the analysis as in the previous study, but we fitted a full model in this case. Salient features of the analysis are: (i) the posterior mixed membership of the 69 students—shown in Figure 7; (ii) the correspondence of latent clusters to student grade levels—shown in Table 1; and (iii) the hyper-parameters were estimated with an empirical Bayes strategy; we obtained $\hat{\alpha} = 0.0487$, $\hat{\rho} = 0.936$, and a practically diagonal matrix that encodes the cluster-to-cluster relations,

$$\hat{B} = \begin{bmatrix} 0.3235 & 0.0 & 0.0 & 0.0 & 0.0 & 0.0 \\ 0.0 & 0.3614 & 0.0002 & 0.0 & 0.0 & 0.0 \\ 0.0 & 0.0 & 0.2607 & 0.0 & 0.0 & 0.0002 \\ 0.0 & 0.0 & 0.0 & 0.3751 & 0.0009 & 0.0 \\ 0.0 & 0.0 & 0.0 & 0.0002 & 0.3795 & 0.0 \\ 0.0 & 0.0 & 0.0 & 0.0 & 0.0 & 0.3719 \end{bmatrix}.$$

4.3   Discussion

There is a tight relationship between ALB and the latent space models in [8,3]. In the latent space models, the latent vectors are drawn from Gaussian distributions and the interaction data is drawn from a Gaussian with mean $\pi_p{}'\mathbb{I}\pi_q$. In ALB, the marginal probability of an interaction takes a similar form, $\pi_p{}'B\pi_q$,

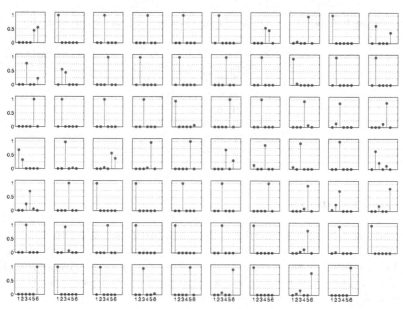

Fig. 7. The posterior mixed membership scores, π, for the 69 students in a school. Each panel correspond to a student; on the Y axis we measure the grade of membership, corresponding to the six grade levels from 7 to 12, on the X axis.

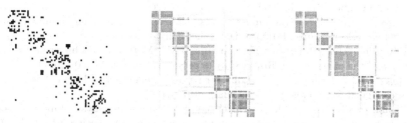

Fig. 8. Original matrix of friensdhip relations (left), and estimated relations obtained by thresholding the posterior expectations $\pi_p{'}B\,\pi_q|R$ (center), and $\phi_p{'}B\,\phi_q|R$ (right)

where B is the matrix of probabilities of interactions for each pair of latent factions. In contrast to the latent space model, the relations can be modeled by an arbitrary distribution, in our model. With binary relations we can use a collection of Bernoulli parameters; with continuous relations, we can use a collection of Gaussian parameters. While more flexible, ALB does not subsume latent space models; they make different assumptions about the data.

Table 1. Grade levels versus (highest) expected posterior membership

Grade	Clusters					
	1	2	3	4	5	6
7	13	1	0	0	0	0
8	0	9	2	0	0	1
9	0	0	16	0	0	0
10	0	0	0	10	0	0
11	0	0	1	0	11	1
12	0	0	0	0	0	4

Acknowledgments

This work was partially supported by National Institutes of Health under Grant No. R01 AG023141-01, by the Office of Naval Research under Contract No. N00014-02-1-0973, by the National Science Foundation under Grants No. DMS-0240019 and DBI-0546594, and by the Department of Defense under Grant No. IIS0218466, all to Carnegie Mellon University.

References

1. Airoldi, E.M., Blei, D.M., Fienberg, S.E., Xing, E.P.: Stochastic block models of mixed membership. Manuscript under review (2006)
2. Sampson, F.S.: A Novitiate in a period of change: An experimental and case study of social relationships. PhD thesis, Cornell University (1968)
3. Handcock, M.S., Raftery, A.E., Tantrum, J.M.: Model-based clustering for social networks. Journal of the Royal Statistical Society, Series A **170** (2007) 1–22
4. Holland, P.W., Leinhardt, S.: Local structure in social networks. In Heise, D., ed.: Sociological Methodology, Jossey-Bass (1975) 1–45
5. Fienberg, S.E., Meyer, M.M., Wasserman, S.: Statistical analysis of multiple sociometric relations. Journal of the American Statistical Association **80** (1985) 51–67
6. Wasserman, S., Pattison, P.: Logit models and logistic regression for social networks: I. an introduction to markov graphs and p^*. Psychometrika **61** (1996) 401–425
7. Snijders, T.A.B.: Markov chain monte carlo estimation of exponential random graph models. Journal of Social Structure (2002)
8. Hoff, P.D., Raftery, A.E., Handcock, M.S.: Latent space approaches to social network analysis. Journal of the American Statistical Association **97** (2002) 1090–1098
9. Doreian, P., Batagelj, V., Ferligoj, A.: Generalized Blockmodeling. Cambridge University Press (2004)
10. Taskar, B., Wong, M.F., Abbeel, P., Koller, D.: Link prediction in relational data. In: Neural Information Processing Systems 15. (2003)
11. Kemp, C., Griffiths, T.L., Tenenbaum, J.B.: Discovering latent classes in relational data. Technical Report AI Memo 2004-019, MIT (2004)
12. Kemp, C., Tenenbaum, J.B., Griffiths, T.L., Yamada, T., Ueda, N.: Learning systems of concepts with an infinite relational model. In: Proceedings of the 21st National Conference on Artificial Intelligence. (2006)

13. McCallum, A., Wang, X., Mohanty, N.: Joint group and topic discovery from relations and text. In: Statistical Network Analysis: Models, Issues and New Directions. Lecture Notes in Computer Science. Springer-Verlag (2007)
14. Blei, D.M., Ng, A., Jordan, M.I.: Latent Dirichlet allocation. Journal of Machine Learning Research **3** (2003) 993–1022
15. Cohn, D., Hofmann, T.: The missing link—A probabilistic model of document content and hypertext connectivity. In: Advances in Neural Information Processing Systems 13. (2001)
16. Erosheva, E.A., Fienberg, S.E., Lafferty, J.: Mixed-membership models of scientific publications. Proceedings of the National Academy of Sciences **97**(22) (2004) 11885–11892
17. Barnard, K., Duygulu, P., de Freitas, N., Forsyth, D., Blei, D., Jordan, M.: Matching words and pictures. Journal of Machine Learning Research **3** (2003) 1107–1135
18. Erosheva, E.A., Fienberg, S.E.: Bayesian mixed membership models for soft clustering and classification. In Weihs, C., Gaul, W., eds.: Classification—The Ubiquitous Challenge. Springer-Verlag (2005) 11–26
19. Manton, K.G., Woodbury, M.A., Tolley, H.D.: Statistical Applications Using Fuzzy Sets. Wiley (1994)
20. Rosenberg, N.A., Pritchard, J.K., Weber, J.L., Cann, H.M., Kidd, K.K., Zhivotovsky, L.A., Feldman, M.W.: Genetic structure of human populations. Science **298** (2002) 2381–2385
21. Pritchard, J., Stephens, M., Donnelly, P.: Inference of population structure using multilocus genotype data. Genetics **155** (2000) 945–959
22. Xing, E.P., Ng, A.Y., Jordan, M.I., Russell, S.: Distance metric learning with applications to clustering with side information. In: Advances in Neural Information Processing Systems. Volume 16. (2003)
23. Holland, P., Laskey, K.B., Leinhardt, S.: Stochastic blockmodels: Some first steps. Social Networks **5** (1983) 109–137
24. Anderson, C.J., Wasserman, S., Faust, K.: Building stochastic blockmodels. Social Networks **14** (1992) 137–161
25. Nowicki, K., Snijders, T.A.B.: Estimation and prediction for stochastic blockstructures. Journal of the American Statistical Association **96** (2001) 1077–1087
26. Airoldi, E.M., Blei, D.M., Fienberg, S.E., Xing, E.P.: Admixtures of latent blocks with application to protein interaction networks. Manuscript under review (2006)
27. Airoldi, E.M., Fienberg, S.E., Xing, E.P.: Latent aspects analysis for gene expression data. Manuscript under review (2006)
28. Carley, K.M.: Smart agents and organizations of the future. In Lievrouw, L., Livingstone, S., eds.: The Handbook of New Media. (2002) 206–220
29. Jordan, M., Ghahramani, Z., Jaakkola, T., Saul, L.: Introduction to variational methods for graphical models. Machine Learning **37** (1999) 183–233
30. Airoldi, E.M., Fienberg, S.E., Joutard, C., Love, T.M.: Discovering latent patterns with hierarchical Bayesian mixed-membership models and the issue of model choice. Technical Report CMU-ML-06-101, School of Computer Science, Carnegie Mellon University (2006)
31. Xing, E.P., Jordan, M.I., Russell, S.: A generalized mean field algorithm for variational inference in exponential families. In: Uncertainty in Artificial Intelligence. Volume 19. (2003)
32. Schervish, M.J.: Theory of Statistics. Springer (1995)
33. Wainwright, M.J., Jordan, M.I.: Graphical models, exponential families and variational inference. Technical Report 649, Department of Statistics, University of California, Berkeley (2003)

34. David, G.B., Carley, K.M.: Clearing the FOG: Fuzzy, overlapping groups for social networks. Manuscript under review (2006)
35. Breiger, R.L., Boorman, S.A., Arabie, P.: An algorithm for clustering relational data with applications to social network analysis and comparison to multidimensional scaling. Journal of Mathematical Psychology **12** (1975) 328–383
36. Harris, K.M., Florey, F., Tabor, J., Bearman, P.S., Jones, J., Udry, R.J.: The national longitudinal study of adolescent health: research design. Technical report, Caorlina Population Center, University of North Carolina, Chapel Hill (2003)
37. Udry, R.J.: The national longitudinal study of adolescent health: (add health) waves i and ii, 1994–1996; wave iii 2001–2002. Technical report, Carolina Population Center, University of North Carolina, Chapel Hill (2003)

Exploratory Study of a New Model
for Evolving Networks

Anna Goldenberg and Alice Zheng

Carnegie Mellon University, Pittsburgh, PA 15213, USA
anya@cs.cmu.edu,alicez@cs.cmu.edu

Abstract. The study of social networks has gained new importance with the recent rise of large on-line communities. Most current approaches focus on deterministic (descriptive) models and are usually restricted to a preset number of people. Moreover, the dynamic aspect is often treated as an addendum to the static model. Taking inspiration from real-life friendship formation patterns, we propose a new generative model of evolving social networks that allows for birth and death of social links and addition of new people. Each person has a distribution over social interaction spheres, which we term "contexts." We study the robustness of our model by examining statistical properties of simulated networks relative to well known properties of real social networks. We discuss the shortcomings of this model and problems that arise during learning. Several extensions are proposed.

1   Introduction

In 1967, the seminal "small world" study [1] brought social networks into the public consciousness. Since then, researchers have paid close attention to laws that seem to govern human and business networks. How do links between people form? Is it enough to look at pairs or should triads of individuals be considered separately? Many approaches study networks on the scale of links and individuals to identify key patterns and describe network properties [2].

Data collection used to be an expensive and tedious process prone to sampling bias. But as more information are becoming available on-line, networks on the order of tens of thousands of people have become easily accessible. Studies of large hyper-link networks reveal similar behavior to those of large social nets (e.g. co-authorships). Thus a new modeling approach has appeared from the random graphs community[3, 4]. Here the goal is not to model the network on a link-by-link basis but to address its overall behavior. The new approach is more generative in nature, though most models are still very simplistic. The preferential attachment model [3] describes the mechanism of network evolution with a focus on power-law degree distributions. Once the links are established, they remain in the network unperturbed. Such simplifying assumptions make the models feasible for analysis, but fail to capture the complexity of real social networks.

E.M. Airoldi et al. (Eds.): ICML 2006 Ws, LNCS 4503, pp. 75–89, 2007.

In this work, we attempt to address several important issues raised by both communities. First, we directly model the generative process behind network dynamics. We focus on the evolution of interpersonal relationships over time, and explicitly model the birth and gradual decay of social links. Secondly, we demonstrate that the model generates networks that exhibit properties commonly observed in many natural topologies.

We motivate our model with an example. Imagine that Andy moves to a new town. He may find some new collaborators at work, make friends at parties, or meet fellow gym-goers while exercising. In general, Andy lives in a number of different spheres of interaction or *contexts*. As time goes on, he may find himself repeatedly meeting certain people in different contexts, consequently developing stronger bonds. Acquaintances he never meets again may quickly fade away. Andy's new friends may also introduce him to their friends (a well known transitive phenomenon called *triadic closures* in social science [2]).

With this example in mind, we begin with a presentation of our model in Section 2. Experimental results are discussed in Section 3. We show how to learn the parameters of our model using Gibbs sampling in Section 4, and give possible extensions of the model in Section 5. Section 6 contains a brief survey of related work, and Section 7 discusses the strengths and weaknesses of the proposed model.

2   The Model

2.1   Notation

DCFM allows the addition of new people into the network at each time step. Let T denote the total number of time steps and N_t the number of people at time t. $N = N_T$ denotes the final total number of people. Let M_t denote the number of new people added to the network at time t, so that $N_t = N_{t-1} + M_t$.

Links between people are weighted. Let $\{W^1, \ldots, W^T\}$ be a sequence of weight matrices, where $W^t \in \mathbb{Z}_+^{N_t \times N_t}$ represents the pairwise link weights at time t. We assume that W^t is symmetric, though it can be easily generalized to the directed case.

The intuition behind our model is that friendships are formed in *contexts*. There are a fixed number of contexts in the world, K, such as work, gym, restaurant, grocery store, etc. Each person has a distribution over these contexts, which can be interpreted as the average percentage of time that he spends in each context.

2.2   The Generative Process

At time t, the N_t people in the network each selects his current context R_i^t from a multinomial distribution with parameter θ_i, where θ_i has a Dirichlet prior distribution:

$$\boldsymbol{\theta}_i \sim \mathrm{Dir}(\boldsymbol{\alpha}), \quad \forall i = 1 : N, \tag{1}$$

$$R_i^t \mid \theta_i \sim \mathrm{Mult}(\theta_i), \quad \forall t = 1 : T, i = 1 : N_t. \tag{2}$$

The number of all possible pairwise meetings at time t is $\text{DYAD}^t = \{(i,j) \mid 1 \leq i \leq N_t, i < j \leq N_t\}$. For each pair of people i and j who are in the same context at time t (i.e., $R_i^t = R_j^t$), we sample a Bernoulli random variable F_{ij}^t with parameter $\beta_i \beta_j$. If $F_{ij}^t = 1$, then i and j meets at time t. The parameter β_i may be interpreted as a measurement of friendliness and is a beta-distributed random variable (making it possible for people to have different levels of friendliness):

$$\beta_i \sim \text{Beta}(a,b), \quad \forall i = 1 : N, \quad \forall (i,j) \in \text{DYAD}^t$$

$$F_{ij}^t \mid R_i^t, R_j^t \sim \begin{cases} \text{Ber}(\beta_i \beta_j) & \text{if } R_i^t = R_j^t \\ I_0 & \text{o.w.} \end{cases} \tag{3}$$

where I_0 is the indicator function for $F_{ij}^t = 0$.

In addition, the newcomers at time t have the opportunity to form triadic closures with existing people. The probability that a newcomer j is introduced to existing person i is proportional to the weight of the links between i and the people whom j meets in his context. Let $\text{TRIAD}^t = \{(i,j) \mid 1 \leq i \leq N_{t-1}, 1 \leq j \leq M_t\}$ denote the pairs of possible triadic closures. For all $(i,j) \in \text{TRIAD}^t$, we have:

$$G_{ij}^t \mid W^{t-1}, F_{\cdot j}^t, R_{\cdot}^t \sim \begin{cases} \text{Ber}(\mu_{ij}^t) & \text{if } R_i \neq R_j \\ I_0 & \text{o.w.,} \end{cases} \tag{4}$$

where $\mu_{ij}^t := \sum_{\ell=1}^{N_t} W_{i\ell}^{t-1} F_{\ell j}^t / \sum_{\ell=1}^{t-1} W_{i\ell}^{t-1}$.

Connection weight updates are Poisson distributed. Our choice of a discrete distribution allows for sparse weight matrices, which are often observed in the real world. Pairwise connection weights may drop to zero if the pair have not interacted for a while (though nothing prevents the connection from reappearing in the future). If i and j meets ($F_{ij}^t = 1$ or $G_{ij}^t = 1$), then W_{ij}^t has a Poisson distribution with mean equal to a multiple (γ_h) of their old connection strength. γ_h signifies the rate of weight increase as a result of the "effectiveness" of a meeting; if $\gamma_h > 1$, then the weight will in general increase. (The weight may also decrease under the Poisson distribution, a consequence perhaps of unhappy meetings.) If i and j do not meet, their mean weight will decrease with rate $\gamma_\ell < 1$. Thus

$$W_{ij}^t \mid W_{ij}^{t-1}, F_{ij}^t, G_{ij}^t, \gamma_h, \gamma_\ell$$

$$\sim \begin{cases} \text{Poi}(\gamma_h(W_{ij}^{t-1} + \epsilon)) & \text{if } F_{ij}^t = 1 \text{ or } G_{ij}^t = 1 \\ \text{Poi}(\gamma_\ell W_{ij}^{t-1}) & \text{o.w.} \end{cases} \tag{5}$$

where $W_{ij}^{t-1} = 0$ by default for $(i,j) \notin \text{TRIAD}^t$, and ϵ is a small positive constant that lifts the Poisson mean away from zero. As W_{ij}^{t-1} becomes large, γ_h and γ_ℓ control the increase and decrease rates, and the effect of ϵ diminishes. γ_h and γ_ℓ have conjugate gamma priors:

$$\gamma_h \sim \text{Gamma}(c_h, d_h), \tag{6}$$

$$\gamma_\ell \sim \text{Gamma}(c_\ell, d_\ell). \tag{7}$$

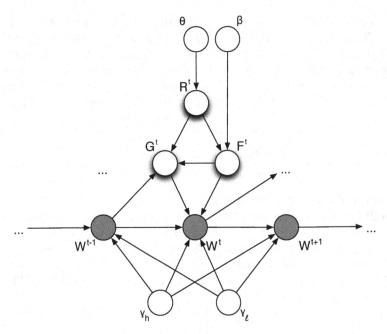

Fig. 1. Graphical representation of one time step of the generative model. R^t is a N_t-dimensional vector indicating each person's context at time t. F^t is a $N_t \times N_t$ matrix indicating pairwise dyadic meetings. G^t is a $N_{t-1} \times M_t$ matrix that indicate triadic closure for newcomers at time t. W^t is the matrix of observed connection weights at time t. θ, β, γ_h, and γ_ℓ are parameters of the model (hyperparameters are not shown).

Figure 1 contains a graphical representation of our model. The complete joint probability is:

$$P(\boldsymbol{\theta}, \boldsymbol{\beta}, \gamma_h, \gamma_\ell, W^{1:T}, R^{1:T}, F^{1:T}, G^{1:T})$$
$$= P(\boldsymbol{\theta})P(\boldsymbol{\beta})P(\gamma_h)P(\gamma_\ell) \prod_t P(R^t|\boldsymbol{\theta})P(F^t|R^t,\boldsymbol{\beta}) \times$$
$$P(G^t|R^t, F^t, W^{t-1})P(W^t|G^t, F^t, W^{t-1}) \quad (8)$$

3   Experiments

We illustrate the behavior of our model under different parameter settings on a set of established metrics.

3.1   Metrics

Degree distribution. In an undirected graph, the degree of a node is its number of neighbors. For node i, we define its degree d_i to be $\sum_{j=1}^{N} I_{(W_{ij}>0)}$, and the average degree of the graph $\sum_{i=1}^{N} d_i/N$.

Node degrees in large natural networks often follow a power law distribution [5], i.e., the number of nodes D with degree n roughly conforms to the function $D(n) = n^{-\rho}$ for some exponent ρ. The value of ρ may vary from network to network, but the overall functional form remains the same. Intuitively, this means that there are many people with a few friends, and very few people with a lot of friends.

Clustering coefficient. Across different social networks, it has often been observed that subsets of people tend to form fully-connected cliques. This inherent clustering tendency may be quantified by the *clustering coefficient* [6]. For the i-th node, C_i is defined to be the ratio between the number of edges E_i that actually exist between its d_i neighbors and the number of edges that would exist if the neighbors form a clique: $C_i = \frac{2E_i}{d_i(d_i-1)}$. The clustering coefficient of the whole network is the average over all nodes: $C = \sum_i C_i/N$.

Average path length. We compute the length of the shortest path s_{ij} between every pair of nodes i and j. If i and j are not connected, then $s_{ij} = \infty$. Let $S := \{(i,j) \mid s_{ij} < \infty\}$ be the set of connected pairs. The average path length of the graph is defined to be $\bar{s} := \sum_{(i,j)\in S} s_{ij}/|S|$.

Effective diameter. The diameter of a graph is the maximum of the shortest path distances between any pair of nodes: $\max_{(i,j)} s_{ij}$. If the graph consists of several disconnected clusters, its diameter is defined to be the maximum over all cluster diameters. Graph diameter can be heavily influenced by outliers. A more robust quantity is the effective diameter, commonly defined as the ninetieth percentile of all shortest paths. Let $\sigma(x)$ be the empirical quantile function of shortest path lengths, i.e., $\sigma(x) = \text{argmax}_s\{s \mid f(s) < x\}$, where $f(s) = |\{(i,j) : s_{ij} < s\}|/N^2$ is the empirical cumulative distribution of s_{ij}. The effective diameter is taken to be $\sigma(.90)$, linearly interpolated if there is no exact match for the ninetieth percentile.

3.2   Simulations

We analyze the behavior of the model under different parameter settings using the four metrics introduced above. [5] and [4] observe a wide range of values for these metrics in a variety of real social networks. Our model can generate networks whose clustering coefficient, average path length, and effective diameter fall within the range of observed values. Here we discuss how different parameter settings affect the values of these metrics, and provide intuition about why this is so.

Unless otherwise specified, the number of contexts K is set to 10. The context preference parameter θ_i is drawn from a peaked Dirichlet prior, where $\alpha_{k^*} = 5$ for a randomly selected k^*, and $\alpha_k = 1$ otherwise. This means that each person in the network has a slight preference for one context. The friendliness parameter β_i is drawn from a Beta(a, b) distribution, where $a = 1$ and b varies. The weights update rates are $\gamma_h = 2$, $\gamma_\ell = 0.5$, and $\epsilon = 1$. We add one person to the network at every time step, so that $n_t = t$. All experiments are repeated with 10 trials.

Friendliness. The parameter β_i determines the "friendliness" of the i-th person and is drawn from a Beta(a, b) distribution. As b increases from 2 to 10, average friendliness decreases from 0.33 to 0.09. We wish to test the effect of b on overall network properties. In order to isolate the effects of friendliness, we fix the context assignments by setting $R_i^t = R_i^1$ for all $t > 1$. In this setting, people do not form triadic closures, and connection weights are updated only through dyadic meetings.

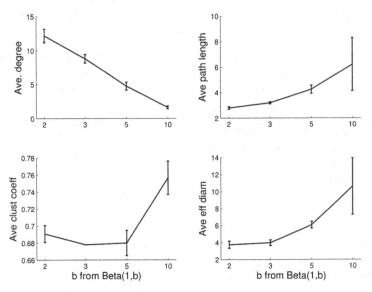

Fig. 2. Effects of the friendliness parameter on a network of 200 people with fixed contexts. The x-axes represent different values of b in Beta$(1, b)$.

As people become less friendly, one expects a corresponding decrease in average node degree. This is indeed what we observe in the average degree plot in Figure 2. Interestingly, the clustering coefficient goes up as friendliness goes down. This is because low friendliness makes for smaller clusters, and it is easier for smaller clusters to become densely connected than it is for bigger clusters. We also observe large variance in average path length and effective diameter at low friendliness levels. This is due to the fact that most clusters now contain one to two people. As small clusters become connected by chance, shortest path lengths varies from trial to trial.

Frequency of Context Switching. In the current model, each person draws a new context at every time step. However, we can easily imagine a person working on one project for a while and then switching to the next project. When context

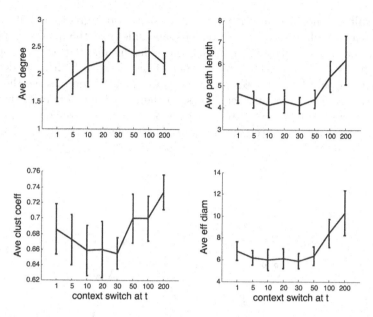

Fig. 3. Effects of the frequency of context switching on a network of 200 people ($b = 8$)

switching is infrequent, people may develop stronger (and more) within-context relations.

We vary the frequency of context switching from 1 to 200 on a 200 node network. When the frequency is 1, people switch context at every time step; when the frequency is 200, contexts are fixed once and for all. In Figure 3, there appears to be a phase transition when context switching occurs every 30 time steps. This occurs as the consequence of two effects. First, when people switch contexts too frequently, they do not have the opportunity to meet everybody in the same context before moving on. Thus they have fewer neighbors and form smaller clusters on average. (As previously discussed, smaller clusters can lead to higher clustering coefficients.) Consequently, the average path length and effective diameter are also slightly long. On the other hand, when people never switch contexts (right-hand end of the x-axes), the number of neighbors is upper bounded by the number of people in the context. Clustering coefficient is high because everybody in the same context knows everybody else, and average path length and diamter are long because there are few paths to people outside of the current context.

Degree Distribution. Under different parameter settings, our model may generate networks with a variety of degree distributions. Lower levels of friendliness typically lead to more power-law-like degree distributions, while higher levels often result in a heavier tail. In Figure 4, we show two degree distribution plots for different friendliness levels. In the left-hand side plot, the quadratic polynomial is a much better fit than the linear one. This means that, when people are more friendly, the drop off in the number of people with high node degree is slower than would be expected under the power law. We do observe the power law effect at a lower level of friendliness. In the right-hand side plot, the linear polynomial with coefficient -1.6 gives as good of a fit as a quadratic function. This coefficient value lies well within the normally observed range for real social networks [5].

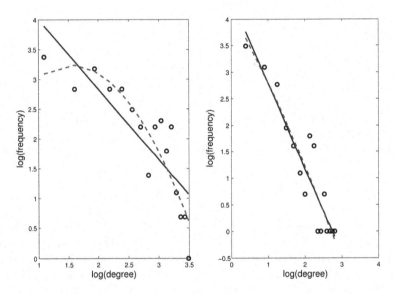

Fig. 4. Log-log plot of the degree distributions of a network with 200 people. β_i is drawn from Beta$(1, 3)$ for the plot on the left, and from Beta$(1, 8)$ for the right hand side. Solid lines represent a linear fit and dashed lines quadratic fit to the data. Contexts are drawn every 50 iterations.

Birth and Death of Links. Our proposed model attempts to capture the dynamics of the birth and death of links. A link is born when the connection weight becomes non-zero, and the link dies when the weight returns to zero. Figure 5 shows link birth rates as the proportion of newly established ties to the number of possible births, and link death rates as the proportion of the number of deaths to the number of links that exist at that point in time.

At the beginning, there are few existing links. Therefore the birth rate is relatively high. Since one person is added to the network at each time step, the number of possible connections grows as $t(t-1)/2$. Thus the birth rate becomes smaller at larger values of t. We note periodical trends in both births and deaths

Birth ratio

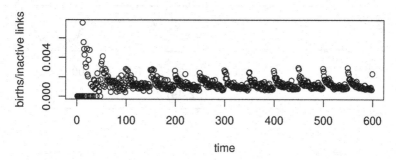

Death ratio

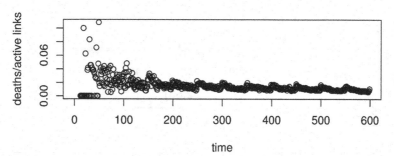

Fig. 5. Birth (top) and death (bottom) of links in a network of 600 people over 600 time steps. Contexts switches occur every 50 iterations, $K = 20$ and $b = 10$.

of links. This periodicity coincides with changes in context. At each context switch, a fresh pool of possible connections becomes available, and weaker links from previous connections are now more likely to die out.

Weight Distributions. One of the main strengths of our model lies in its ability to represent weighted links. In real life, friendships are not simply existent or absent. A strong connection should take longer to dissipate than would a weak connection. Link weights act as memory in preserving friendships. Old friendships may be rekindled if the pair rotate within similar contexts. We compare the evolution of simulated weights with email exchange in the well-known Enron dataset. Figure 6 shows typical weight progressions over time in a simulated network. Figure 7 plots typical patterns of weekly email exchange counts between Enron employees. Our model is clearly capable of reproducing both long-lasting and short-range connections. Previously severed links can be renewed, as is the case for the pair $(45, 47)$.

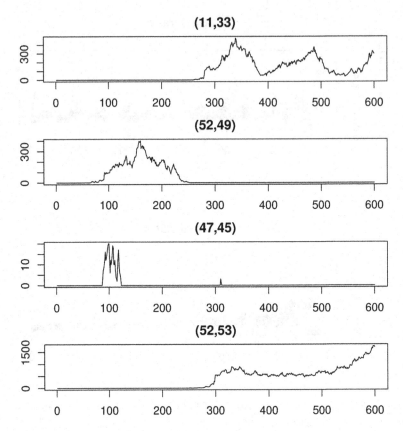

Fig. 6. Weight dynamics for 4 different pairs in a network of 600 people over 600 time steps. Contexts switches occur every 50 iterations and $b = 3$.

4   Learning Parameters Via Gibbs Sampling

Parameter learning in DCFM is possible via Gibbs sampling. We leave a detailed investigation of learning results to another paper, but give the Gibbs updates here for reference. Using ... as a shorthand for "all other variables in the model," we have:

$$\boldsymbol{\theta}_i \mid \ldots \sim \text{Dir}(\boldsymbol{\alpha} + \boldsymbol{\alpha}_i'), \tag{9}$$

$$P(\beta_i \mid \ldots) \propto \beta_i^{A_i + a - 1} (1 - \beta_i)^{b-1} \prod_{j \neq i} (1 - \beta_i \beta_j)^{B_{ij}}, \tag{10}$$

$$\gamma_h \mid \ldots \sim \text{Gamma}(c_h + w_h, (v_h + 1/d_h)^{-1}), \tag{11}$$

$$\gamma_\ell \mid \ldots \sim \text{Gamma}(c_\ell + w_\ell, (v_\ell + 1/d_\ell)^{-1}). \tag{12}$$

In Equation 9, $\alpha_{ik}' := \sum_{t=1}^{T} I_{(R_i = k)}$ is the total number of times person i is seen in context k. In Equation 10, $A_i := |\{(j, t) \mid R_i^t = R_j^t \text{ and } F_{ij}^t = 1\}|$

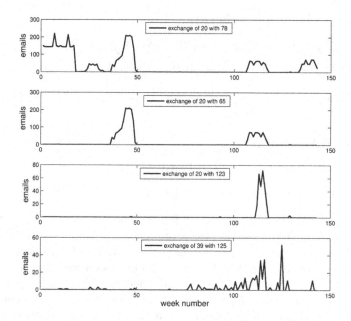

Fig. 7. Weekly email exchange counts for four randomly selected pairs between 136 Enron employees

is the total number of dyadic meetings between i and any other person, and $B_{ij} := |\{t \mid R_i^t = R_j^t \text{ and } F_{ij}^t = 0\}|$ is the total number of times i has "missed" an opportunity for a dyadic meeting. Let $H := \{(i,j,t) \mid F_{ij}^t = 1 \text{ or } G_{ij} = 1\}$ represent the union of the set of dyadic and triadic meetings, and $\mathcal{L} := \{(i,j,t) \mid (i,j) \in \text{DYAD}^t \text{ and } F_{ij}^t = 0\}$ the set of missed dyadic meeting opportunities. $w_h := \sum_{(i,j,t) \in \mathcal{H}} W_{ij}^t$ is the sum of updated weights after the meetings, and $v_h := \sum_{(i,j,t) \in \mathcal{H}} (W_{ij}^{t-1} + \epsilon)$ is the sum of the original weights plus a fixed constant. $w_l := \sum_{(i,j,t) \in \mathcal{L}} W_{ij}^t$ is the sum of weights after the missed meetings, and $v_l := \sum_{(i,j,t) \in \mathcal{L}} W_{ij}^{t-1}$ is the sum of original weights. (Here we use zero as the default value for W_{ij}^{t-1} if j is not yet present in the network at time $t-1$.)

Due to coupling from the pairwise interaction terms $\beta_i \beta_j$, the posterior probability distribution of β_i cannot be written in a closed form. However, since β_i lies in the range $[0,1]$, one can perform coarse-scale numerical integration and sample from interpolated histograms. Alternatively, one can design Metropolis-Hasting updates for β_i, which has the advantage of maintaining a proper Markov chain.

The variables F_{ij}^t and G_{ij} are conditionally dependent given the observed weight matrices. If a pairwise connection W_{ij} increases from zero to a positive value at time t, then i and j must either have a dyadic or a triadic meeting. On the other hand, dyadic meetings are possible only when i and j are in the same context, and triadic meetings when they are in different contexts. Hence F_{ij}^t and G_{ij}^t may never both be 1. In order to ensure consistency, F_{ij}^t and G_{ij} must be updated together. For $(i, j) \in \text{TRIAD}^t$,

$$P(F_{ij}^t = 1, G_{ij} = 0 \mid \ldots) \propto I_{(R_i^t = R_j^t)}(\beta_i \beta_j) \text{Poi}(W_{ij}^t; \gamma_h \epsilon),$$

$$P(F_{ij}^t = 0, G_{ij} = 1 \mid \ldots) \propto I_{(R_i^t \neq R_j^t)} \mu_{ij} \text{Poi}(W_{ij}^t; \gamma_h \epsilon),$$

$$P(F_{ij}^t = 0, G_{ij} = 0 \mid \ldots) \propto \left[I_{(R_i^t = R_j^t)}(1 - \beta_i \beta_j) + I_{(R_i^t \neq R_j^t)}(1 - \mu_{ij}) \right] I_{(W_{ij}^t = 0)}.$$

$$(13)$$

For $(i, j) \in \text{DYAD}^t \setminus \text{TRIAD}^t$,

$$P(F_{ij}^t = 1 \mid \ldots) \propto I_{(R_i^t = R_j^t)}(\beta_i \beta_j) \text{Poi}(W_{ij}^t; \gamma_h (W_{ij}^{t-1} + \epsilon)),$$

$$P(F_{ij}^t = 0 \mid \ldots) \propto (I_{(R_i^t = R_j^t)}(1 - \beta_i \beta_j) + I_{(R_i^t \neq R_j^t)}) \text{Poi}(W_{ij}^t; \gamma_\ell W_{ij}^{t-1}).$$

$$(14)$$

There are also consistency constraints for R^t. For example, if $F_{ij}^t = F_{jk}^t = 1$, then i, j, and k must all lie within the same context. If $G_{kl} = 1$ in addition, then l must belong to a different context from i, j, and k. The F variables propagate transitivity constraints, whereas G propagates exclusion constraints.

To update R^t, we first find connected components within F^t. Let p denote the number of components and I the index set for the nodes in the i-th component. We update each R_I^t as a block. Imagine an auxiliary graph where nodes represent these connected components and edges represent exclusion constraints specified by G, i.e., I is connected to J if $G_{ij} = 1$ for some $i \in I$ and $j \in J$. Finding a consistent setting for R^t is equivalent to finding a feasible K-coloring of the auxiliary graph, where K is the total number of contexts. We sample R_I^t sequentially according to an arbitrary ordering of the components. Let $\pi(I)$ denote the set of components that are updated before I. The posterior probabilities are:

$$P(R_I^t = k \mid R_{\pi(I)}^t, G) \propto \begin{cases} 0 & \text{if } G_{IJ} = 1 \text{ and } R_J^t = k \text{ for some } J \in \pi(I) \\ \prod_{i \in I} \theta_{ik} & \text{otherwise} \end{cases}$$

$$(15)$$

These sequential updates correspond to a greedy K-coloring algorithm; they are approximate Gibbs sampling steps in the sense that they do not condition on the entire set of connected components.

5   Possible Extensions

5.1   Evolution of Context Preferences

A person's context distribution is influenced by the social groups to which he belongs. People who are friends with gym-goers may start to frequent the gym themselves. Thus it could be desirable to incorporate evolution of the θ parameters (indicating context preference) into our model. We propose to update θ for each person using the θ parameters of his neighbors, weighted by the connection strengths:

$$\theta_i^t = \lambda \theta_i^{t-1} + (1 - \lambda) \frac{1}{\sum_j W_{ij}^t} \sum_j W_{ij}^t \theta_j^{t-1}.$$

$$(16)$$

The larger λ (a person's independence) is, the less susceptible the person is to the preference of his friends.

5.2 Long Term Memory

Weighted links capture the effect of short term memory; in our model, a link established at time t will likely remain at time $t + 1$. However, once the weight becomes zero, renewal of the link becomes is likely as a 'birth' of a new link. To capture long term memory, we could model weights as a continuous gamma distribution, so that established links always carry small residual weights. The drawback is that the weight matrices will be dense, and we would need an additional thresholding parameter for the 'death' of a link. Alternatively, at the cost of introducing N new parameters, we can make each person 'remember' the strength and duration of his past connections.

6   Related Work

The principles underlying the mechanisms by which relationships evolve are still not well understood [7]. Current models aim at either describing observed phenomena or predicting future trends. A common approach is to select a set of graph based features, such as degree distribution or the number of dyads and triangles, and create models that mimic observed behavior of the evolution of these features in real life networks. Works [8, 9, 10] in physics and [11, 12] in social sciences follow this approach. However, under models of average behavior, the actual links between any two given people might not have any meaning. Consequently, these models are often difficult to interpret.

Another approach aims to predict future friends and collaborators based on the properties of the network seen so far [4, 7]. These models often cannot encode common network dynamics such as mobility and link modification. Moreover, these models usually do not take into account triadic closure, a phenomenon of great importance in social networks [2, 13].

[14] presents an interesting dynamic social network model (with fixed number of people). This work builds on [15], which introduces latent positions for each person in order to explain observed links. If two people are close in the latent space, they are likely to have a connection. [15] estimate latent positions in a static data set. [14] adds a dynamic component by allowing the latent positions to be updated based on both their previous positions and on the newly observed interactions. One can imagine a generative mechanism that governs such perturbations of latent positions. In fact, the DCFM model presented in this paper can be seen as a generative model for the latent mapping function.

7   Discussion

Our focus on generative modeling in this paper is prompted by the need to provide a plausible explanation for how networks form and evolve. It is flexible and can be adapted to alternative theories of the friend evolution process. For example, in our model, the decision to allow links to decay is made independently on each pair. However, theory of Simmelian ties [16] suggest that two people who are no longer friends may nevertheless remain so due to influence from a third party. This is a plausible alternative to our current model.

Our choice of modeling weighted networks is motivated by the fact that friednships between people are not binary. Stronger links tend to last longer periods of time; temporary connections cease to exist once the cause disappears. However, it is often difficult to obtain real datasets with weighted connections. We propose to use the number of email, sms and phone call exchanges in preset time intervals as a proxy to the weight of links between people. This is a very coarse representation of a relationship weight, since non-communication does not necessarily imply change in link weight. Hence the DCFM model may predict smoother connection weights than the observed values.

To show that our model is capable of generating realistic social environments, we provide simulation results that adhere to observations made on realistic datasets in [17]. However, there is no groundtruth for the parameters in the hidden layer. Variables that address context choice and meeting occurrance at time step t have to be inferred from the previous and currently observed weights alone. This brings up the question of identifiability. Unfortunately, the complexity of the model makes it difficult to answer this question and we are currently exploring possible solutions to this problem.

Another interesting question is exchangeability. The earlier a person appears in the network, the more chances he has to establish connections. People who have been in the network longer are expected to have more connections and thus nodes (people) are not exchangeable over time.

The current model does not place any explicit upper bounds on the number of links a person can establish. It is effectively limited by the number of people in the same context. Unless a person is very friendly and has uniform distribution, the number of links is not expected to be high. In realistic networks, we expect the context preference distribution and friendliness to be skewed, because a person has a limited amount of time and energy to build and maintain relationships.

In conclusion, we provide an exploratory study of a new generative model for dynamic social networks in this paper. Simulation results demonstrate the advantages as well as shortcomings of this model. In future work, we hope to address issues of identifiability and investigate possible extensions of this work.

References

[1] Milgram, S.: The small-world problem. Psychology Today (1967) 60–67
[2] Wasserman, S., Faust, K.: Social Network Analysis: Methods and Applications. Cambridge University Press, Cambridge (1994)
[3] Barabási, A.L., Albert, R.: Emergence of scaling in random networks. Science **286** (1999) 509–512
[4] Newman, M.: The structure of scientific collaboration networks. In: Proceedings of the National Academy of Sciences USA. Volume 98. (2001) 404–409

[5] Albert, R., Barabási, A.: Statistical mechanics of social networks. Rev of Modern Physics **74** (2002) 47–97

[6] Watts, D., Strogatz, S.: Collective dynamics of "small-world" networks. Nature **393** (1998) 440–442

[7] Liben-Nowell, D., Kleinberg, J.: The link prediction problem for social networks. In: Proc. 12th International Conference on Information and Knowledge Management. (2003)

[8] Jin, E., Girvan, M., Newman, M.: The structure of growing social networks. Physical Review Letters E **64** (2001) 381–399

[9] Barabasi, A., Jeong, H., Neda, Z., Ravasz, E., Schubert, A., Vicsek, T.: Evolution of the social network of scientific collaboration. Physica A **311**(3–4) (2002) 590–614

[10] Davidsen, J., Ebel, J., Bornholdt, S.: Emergence of a small world from local interactions: Modeling acquaintance networks. Physical Review Letters **88** (2002) 128701-1–12870-4

[11] Van De Bunt, G., Duijin, M.V., Snijders, T.: Friendship networks through time: An actor-oriented dynamic statistical network model. Computation and Mathematical Organization Theory **5**(2) (1999) 167–192

[12] Huisman, M., Snijders, T.: Statistical analysis of longitudinal network data with changing composition. Sociological Methods and Research **32**(2) (2003) 253–287

[13] Kossinets, G., Watts, D.: Empirical analysis of an evolving social network. Science **311**(5757) (2006) 88–90

[14] Sarkar, P., Moore, A.: Dynamic social network analysis using latent space models. SIGKDD Explorations: Special Edition on Link Mining (2005)

[15] Hoff, P., Raftery, A., Handcock, M.: Latent space approaches to social network analysis. Journal of the American Statistical Association **97** (2002) 1090–1098

[16] Krackhardt, D.: The ties that torture: Simmelian tie analysis in organizations. Research in the Sociology of Organizations **16** (1999) 183–210

[17] Albert, R., Barabási, A.L.: Dynamics of complex systems: Scaling laws for the period of boolean networks. Physical Review Letters **84** (2000) 5660–5663

A Latent Space Model for Rank Data

Isobel Claire Gormley and Thomas Brendan Murphy

Department of Statistics, School of Computer Science and Statistics,
Trinity College Dublin, Dublin 2, Ireland
{gormleyi, murphybt}@tcd.ie

Abstract. Rank data consist of ordered lists of objects. A particular example of these data arises in Irish elections using the proportional representation by means of a single transferable vote (PR-STV) system, where voters list candidates in order of preference. A latent space model is proposed for rank (voting) data, where both voters and candidates are located in the same D dimensional latent space. The relative proximity of candidates to a voter determines the probability of a voter giving high preferences to a candidate. The votes are modelled using a Plackett-Luce model which allows for the ranked nature of the data to be modelled directly. Data from the 2002 Irish general election are analyzed using the proposed model which is fitted in a Bayesian framework. The estimated candidate positions suggest that the party politics play an important role in this election. Methods for choosing D, the dimensionality of the latent space, are discussed and models with $D = 1$ or $D = 2$ are proposed for the 2002 Irish general election data.

1 Introduction

Proportional representation by means of a single transferable vote (PR-STV) is the electoral system employed in Ireland in both general (governmental) and presidential elections. In this electoral system, voters are required to rank some or all of the proposed candidates in order of preference.

A wealth of rank data is available within the context of Irish elections. The introduction of electronic voting in several constituencies means actual voting data are now publicly available. The work presented here focuses on the Irish general election of 2002, where the current government was elected; specifically the votes from the general election in the constituency of Meath are examined. Details of this election are outlined in Section 2.

A latent space model (Section 3.1) similar to that of Hoff et al. [1] is proposed where both voters and candidates are located simultaneously in a D-dimensional latent space. The location of each candidate is inferred from the votes cast by the electorate — the Plackett-Luce model for rank data (Section 3.2) is employed to exploit the information incorporated in the ranked preferences contained in the votes. In turn, voter locations are determined by their votes, which demonstrate their support for each of the candidates. This model is fitted within the Bayesian paradigm; the Metropolis-Hastings algorithm is the primary model fitting tool.

E.M. Airoldi et al. (Eds.): ICML 2006 Ws, LNCS 4503, pp. 90–102, 2007.

When fitting latent space models issues such as invariant configurations and choice of dimensionality arise; these are dealt with in Sections 4.2 and 4.3 respectively.

The relative spatial locations of the candidates are suggestive of the types of relationships that may exist between the candidates, as viewed by the electorate. As coalition governments often occur in countries that use proportional representation election systems, interest lies in examining if candidates from different political parties are deemed alike. Which political parties are viewed as similar by the electorate? What characteristics do closely located candidates share? What mechanisms drive Irish general elections? Such questions will be answered by examining the relative locations of the candidates.

Configurations of the candidates and electorate from the 2002 general election in the Meath constituency indicate that party politics drive voter opinions.

We conclude, in Section 6, by proposing possible modifications and extensions to the model fitted in this study.

2   Irish Elections

Irish elections employ a voting system called proportional representation by means of a single transferable vote (PR-STV). In the PR-STV system, voters rank some or all of the candidates in order of preference. During the counting process first preferences are totalled and candidates are elected or eliminated from the race depending on a constituency specific quota. Excess votes above the quota or those belonging to an eliminated candidate are transferred to other candidates according to the ranked preferences. This process continues until all seats are filled or until a sufficient number of candidates are left in the race. A precise description of the electoral system, including the method of counting votes is given by Sinnott [2]. The transfer of votes during the 2002 general election in the Meath constituency can be viewed at http://www.oireachtas.ie. Good introductions to the Irish political system are given in Coakley and Gallagher [3] and Sinnott [4].

2.1   The 2002 General Election

Dáil Éireann (the Irish House of Parliament) consists of one hundred and sixty six members who are elected in a general election held at least once every five years. The members of the Dáil represent forty two constituencies. The most recent general election was held on May 17th, 2002. This election saw the introduction of electronic voting, for the first time, in three constituencies (Dublin North, Dublin West, and Meath). The remaining thirty nine constituencies used paper ballots.

Five seats in Dáil Éireann were allocated to the constituency of Meath and fourteen candidates ran for election within the constituency. The fourteen candidates represented seven political parties, with the major parties of Fianna Fáil and Fine Gael each having three candidates. Table 1 provides details of all the candidates and their political affiliations. The five seats in the Meath constituency were won by Dempsey, Bruton, Wallace, English and Brady. The

Table 1. The fourteen candidates who ran for election in the Meath constituency in 2002. The political affiliation of each candidate is given as well as an abbreviation of their surname and party. Independent candidates are not affiliated to any party. The candidates shown in boldface were elected.

Candidate	Party
Brady, J. (By)	Fianna Fáil (FF)
Bruton, J. (Bt)	Fine Gael (FG)
Colwell, J. (Cl)	Independent (Ind)
Dempsey, N. (Dp)	Fianna Fáil (FF)
English, D. (Eg)	Fine Gael (FG)
Farrelly, J. (Fr)	Fine Gael (FG)
Fitzgerald, B. (Fz)	Independent (Ind)
Kelly, T. (Kl)	Independent (Ind)
O'Brien, P. (Ob)	Independent (Ind)
O'Byrne, F. (Oby)	Green Party (GP)
Redmond, M. (Rd)	Christian Solidarity (CSP)
Reilly, J. (Rl)	Sinn Féin (SF)
Wallace, M. (Wl)	Fianna Fáil (FF)
Ward, P. (Wd)	Labour (Lab)

electorate in Meath consisted of 108,717 individuals and there were a total of 64,081 valid votes cast. The actual votes cast in the Meath constituency are analyzed in this work.

The voting data from the Meath constituency is available from the Meath local authority web page (`http://www.meath.ie/election.html`). The voting data from the other constituencies where electronic voting was implemented is available from (`http://www.dublincountyreturningofficer.com`).

3 Modeling

A latent space model is combined with a model for rank data to provide a suitable tool for the modeling of PR-STV data.

3.1 The Latent Space Model

Hoff et al. [1] proposed a model for social networks where the network actors are located in a latent space and the probability of a connection between two actors is determined by their proximity. In a similar vein to this work, a model is proposed for rank data where voters and candidates are located in the same D dimensional latent space $\mathbf{Z} \subseteq \Re^D$. It is assumed that each of M voters has latent location $\underline{z}_i \in \mathbf{Z}$ and each candidate j ($j = 1, \ldots, N$) has latent location $\underline{\zeta}_j \in \mathbf{Z}$. Hence, the N preferences of M voters are described using $(N+M)D$ parameters.

Let $d(\underline{z}_i, \underline{\zeta}_j)$ be the squared Euclidean distance between voter i and candidate j in the latent space $\mathbf{Z}$, that is

$$d(\underline{z}_i, \underline{\zeta}_j) = \frac{1}{D} \sum_{d=1}^{D} (z_{id} - \zeta_{jd})^2$$

for $1 = 1, \ldots, M$ and $j = 1, \ldots, N$. The squared Euclidean distance is invariant to rotations and translations. Many other distance measures are available [5] and possible alternatives are discussed in Section 6.

The distance $d(\underline{z}_i, \underline{\zeta}_j)$ (for $j = 1 \ldots, N$) between voter i and the candidates describes the voter's electoral opinions. In a similar way the proximity of two candidates in the latent space quantitatively describes their relationship as deemed by the electorate.

By exploiting the information contained in the ranked preferences the latent locations of each voter and candidate can be inferred. Thus a latent space model is incorporated with a standard rank data model to spatially model Irish voting data.

3.2   The Plackett-Luce Model

In the context of PR-STV voting, the data value $\mathbf{x}_i = (c(i, 1), c(i, 2), \ldots, c(i, n_i))$ denotes voter i's ballot, where n_i is the number of preferences expressed by voter i and $c(i, t)$ denotes the candidate chosen in tth position by voter i. Under the PR-STV voting system the number of preferences voter i expresses may vary; that is $1 \leq n_i \leq N$.

Many rank data models in the literature emerge from the context of predicting the final ordering of horses in a race (see [6, 7]). The ranking of candidates on a ballot form may be thought of in a similar manner. Marden [8] details many of the currently available rank data models. The Plackett-Luce model [6] is one such model which can be used to model PR-STV data. Mixtures of Plackett-Luce models have recently been used by Gormley and Murphy [9] to model Irish college applications data.

In the Plackett-Luce model, a ranking is modeled as a process in which each voter sequentially selects the next most preferred candidate; such sequential models are called *multi stage models* [10]. The model is parameterized by a 'support' parameter vector

$$\underline{p}_i = (p_{i1}, p_{i2}, \ldots, p_{iN}),$$

where $p_{ij} = \mathbf{P}\{$Voter i ranking candidate j first|Available candidates$\}$. The model assumes that $\mathbf{P}\{$Voter i ranking candate j in position t|Available candidates$\} \propto p_{ij}$. Hence, the probability of vote $\mathbf{x}_i = (c(i, 1), c(i, 2), \ldots, c(i, n_i))$ is

$$\mathbf{P}\{\mathbf{x}_i | \underline{p}_i\}$$

$$= \prod_{t=1}^{n_i} \mathbf{P}\{\text{Voter } i \text{ ranking candidate } c(i, t) \text{ in position } t | \text{Available candidates}\}$$

$$= \prod_{t=1}^{n_i} \frac{p_{ic(i,t)}}{\sum_{s=t}^{N} p_{ic(i,s)}}$$

where for $s > n_i$ the sequence of $c(i, s)$ is any arbitrary ordering of the candidates not selected by voter i; this ordering does not affect any calculations because these values only arise in calculations involving $\sum_{s=n_i+1}^{N} p_{ic(i,s)}$ which does not depend on the ordering.

We assume that probability p_{ij} is a decreasing function of the distance between the voter and the candidate in the latent space. It is assumed that these probabilities take the form

$$p_{ij} = \frac{\exp\{-d(\underline{z}_i, \underline{\zeta}_j)\}}{\sum_{j'=1}^{N} \exp\{-d(\underline{z}_i, \underline{\zeta}_{j'})\}}$$

for $i = 1, \ldots, M$ and $j = 1, \ldots, N$. Thus, the probability of vote $\mathbf{x}_i$ is determined by the location of the voter and candidates in the latent space.

4   Model Fitting

The Plackett-Luce model combined with a latent space model allows for the ranked nature of the PR-STV data to be spatially modeled. The parameters of this model and their related uncertainty are estimated within a Bayesian framework.

Prior densities for voter locations, $p_v(\underline{z}_i)$, and for candidate locations, $p_c(\underline{\zeta}_j)$ are assumed to be Normal and independent where $z_{id} \sim N(0, 3^2)$ and $\zeta_{jd} \sim N(0, 3^2)$ for $d = 1, \ldots, D$. The prior parameters were selected so that the prior was concentrated on a region around the origin without being overly informative. Thus the joint density $P\{\mathbf{X}, \mathbf{z}, \zeta\}$ of the votes cast, the voter locations and the candidate locations is

$$\prod_{i=1}^{M} p_v(\underline{z}_i) \left[\prod_{t=1}^{n_i} \frac{\exp\{-d(\underline{z}_i, \underline{\zeta}_{c(i,t)})\}}{\sum_{s=t}^{N} \exp\{-d(\underline{z}_i, \underline{\zeta}_{c(i,s)})\}} \right] \left[\prod_{j=1}^{N} p_c(\underline{\zeta}_j) \right].$$

The location of each voter and each candidate in the latent space are to be estimated — samples from the posterior distribution $\mathbf{P}\{\mathbf{z}, \zeta | \mathbf{X}\}$ are generated using a Metropolis-Hastings algorithm. A random walk proposal density where each location was perturbed using normally distributed noise was employed — good acceptance rates (detailed in Section 5) were achieved in the estimation of both voter and candidate locations using this proposal.

4.1   Estimation of Voter and Candidate Latent Locations

The location of each voter $\underline{z}_i$ and each candidate $\underline{\zeta}_j$ within a D dimensional latent space is to be estimated. A random walk Metropolis-Hastings algorithm is used to sample from the joint density $\mathbf{P}\{\mathbf{z}, \zeta | \mathbf{X}\}$.

Estimates of $\underline{z}_i$ are generated from the posterior distribution via the following algorithm:

1. Generate a value ϵ from the symmetric proposal density $N(0, \sigma_v^2)$ and form the proposal point $z_{id}^* = z_{id} + \epsilon$ for $d = 1, \ldots, D$.
2. Compute the acceptance probability $\alpha(\underline{z}_i^*, \underline{z}_i)$ as follows

$$\alpha(\underline{z}_i^*, \underline{z}_i) = \min \left\{ \frac{P(\mathbf{z}^*, \zeta | \mathbf{X})}{P(\mathbf{z}, \zeta | \mathbf{X})} \; , \; 1 \right\}$$

where independence of voter locations and a symmetric random walk proposal distribution are assumed.
3. Generate a value $u \sim \text{Uniform}(0, 1)$.
4. If $u \leq \alpha(\underline{z}_i^*, \underline{z}_i)$ then define $\underline{z}_i = \underline{z}_i^*$, otherwise define $\underline{z}_i = \underline{z}_i$.

Similar methodology applies in the case of estimating candidate locations:

1. Generate a value $\epsilon \sim N(0, \sigma_c^2)$ and let $\zeta_{jd}^* = \zeta_{jd} + \epsilon$ for $d = 1, \ldots, D$.
2. Compute the acceptance probability $\alpha(\underline{\zeta}_j^*, \underline{\zeta}_j)$ as follows

$$\alpha(\underline{\zeta}_j^*, \underline{\zeta}_j) = \min \left\{ \frac{P\{\mathbf{z}, \zeta^* | \mathbf{X}\}}{P\{\mathbf{z}, \zeta | \mathbf{X}\}} \; , \; 1 \right\}.$$

where independence of candidates locations and symmetric random walk proposal are assumed.
3. Generate a value $u \sim \text{Uniform}(0, 1)$.
4. If $u \leq \alpha(\underline{\zeta}_j^*, \underline{\zeta}_j)$ then define $\underline{\zeta}_j = \underline{\zeta}_j^*$, otherwise define $\underline{\zeta}_j = \underline{\zeta}_j$.

The algorithm sequentially estimates the voter locations and then the candidate locations until sufficient mixing of the Markov chain is achieved. Locations estimated subsequent to a burn-in period are considered when calculating final estimates.

4.2   Invariant Configurations

The measure of distance between voter and candidate locations in the latent space is quantified by the squared Euclidean distance. This distance is invariant to rotations and translations. As a result the model is not fully identifiable because the locations are only identified up to rotation and translation. Procrustean methods are used to eradicate this problem.

Procrustean methods [11] match one configuration of points to another as well as possible in a least squares sense. Transformations such as dilation, rotation and translation are used to create the match. In this context only translations and rotations are applicable without altering the likelihood of the data due to the definition of the probabilities p_{ij}.

Assume $C^R = (\mathbf{z}^R, \zeta^R)$ is a reference configuration of the voter and candidate locations which is centered around the origin. To match the estimated configuration $\hat{C}$ to the reference configuration C^R, $\hat{C}$ is first translated so that it is also centered around the origin. $\hat{C}$ is then rotated to provide the best match with C^R in a least squares sense.

To obtain Q, the optimal orthogonal rotation matrix, the sum

$$
\begin{aligned}
S &= \sum_{i=1}^{M+N} \sum_{d=1}^{D} (c_{id}^R - \hat{c}_{id})^2 \\
&= \sum_{i=1}^{M} \sum_{d=1}^{D} (z_{id}^R - \hat{z}_{id})^2 + \sum_{j=1}^{N} \sum_{d=1}^{D} (\zeta_{jd}^R - \hat{\zeta}_{jd})^2 \\
&= \text{trace} \left\{ C^R C^{R'} + \hat{C}\hat{C}' - 2C^R \hat{C}' \right\}
\end{aligned}
$$

is minimized. The newly rotated configuration is denoted $\hat{C}Q$. Thus (1) becomes

$$
S = \text{trace} \left\{ C^R C^{R'} + \hat{C}\hat{C}' - 2C^R Q' \hat{C}' \right\}
$$

and the minimization problem becomes the constrained maximization of $2C^R Q' \hat{C}'$. It follows that $Q = VU'$ where $U \Sigma V'$ is the singular value decomposition of $C^{R'} \hat{C}$. Thus by centering each estimated configuration around the origin and rotating the configuration using the rotation matrix Q the estimated configuration $\hat{C}$ is best matched with the reference configuration C^R.

Samples of the configuration $(\mathbf{z}, \zeta)$ are generated using the Metropolis-Hastings algorithm (Section 4.1). Initial iterations of the algorithm are constrained to only accept uphill moves (ie. moves when $\alpha(\underline{z}_i^*, \underline{z}_i) \geq 1$ and $\alpha(\underline{\zeta}_i^*, \underline{\zeta}_i) \geq 1$ which guarantees that the posterior density increases or stays constant on each iteration) to achieve an estimate of the *maximum a posteriori* (MAP) configuration of candidate and voter locations. This MAP configuration is henceforth employed as C^R, the reference configuration, to which each subsequently estimated configuration $\hat{C}$ is matched. C^R is not assumed to be the correct configuration but is merely used as a standard to which others are matched. Locations estimated during the uphill only runs of the Metropolis-Hastings algorithm are not considered when calculating final estimates.

4.3   Dimensionality

The dimensionality D of the latent space is a further variable which requires estimation. Several techniques have been discussed in the literature (eg. [12]) as potential methods of selecting the optimal dimensionality of a space. Selecting the optimal D can been viewed as a model selection process between models with different dimensions.

The deviance information criterion (DIC) [13] is a measure of model complexity defined as a Bayesian measure of accuracy (or deviance) penalized by an additional complexity term. The complexity term is labeled the 'effective number of parameters' and is the difference between the posterior mean of the deviance and the deviance at the posterior mean of the estimates of the model parameters. Spiegelhalter et al. [13] tentatively suggest that combining the Bayesian deviance and the complexity term forms a criterion which may be used for comparing models.

Pritchard et al. [14] suggest a model selection criterion based on an approximation of the posterior distribution $\mathbf{P}\{D|\mathbf{X}\}$. It is computationally similar to the DIC but penalizes the mean of the Bayesian deviance by a quarter of it's variance as opposed to the effective number of parameters.

A practical alternative to these two criteria is to apply principal components analysis [5] to the resulting configurations for each choice of dimension D. The principal components analysis (PCA) rotates the configuration of candidate locations so that the variance of the locations is concentrated in a subset of the dimensions: the first principal component dimension has maximal variance, the second has maximal variance subject to being orthogonal to the first dimension, etc.

PCA is applied to the configuration of candidates only as the predominant interest lies in the interpretation of the relative locations of the candidates. The variances of the resulting principal components are examined and the optimal number of dimensions D is selected to be the number of dimensions after which the addition of another dimension was not deemed to be beneficial; a threshold of 10% was used to determine if the addition of an extra dimension was worthwhile.

5   Results

A latent space model for rank data was applied to the set of votes from the Meath constituency in the 2002 Irish general election.

5.1   The 2002 General Election: Meath Constituency

Five seats in Dáil Éireann were available for election in the Meath constituency in the 2002 general election. A latent space model, incorporating the Plackett-Luce model for rank data, was fitted to the 64,081 electronic votes over the range of dimensions $D = 1, 2$ and 3. The random walk proposal densities were fixed to be

$$N(0, \sigma_v^2) = N(0, 3^2)$$
$$N(0, \sigma_c^2) = N(0, 0.005^2).$$

The DIC and Pritchard et al's criterion were computed (Table 2) to determine the appropriate dimension for the latent space but DIC suggested $D = 3$ whereas Prtichard et al's criterion suggested $D = 1$. Airoldi et al. [12] provide a detailed review of model choice in Bayesian models and emphasize its importance, in particular the problems with fitting overly complex models.

As a result of the conflict between the results using DIC and Pritchard's criterion and the concern about fitting overly complex models, we also used a principal components analysis of the candidate locations to select D. Table 3 shows the variation captured by each principal component when different dimensions of latent space model were fitted to the data. Both dimensions $D = 1$

Table 2. The DIC values and Pritchard et al.'s criterion values for latent space models of dimension $D = 1, 2$ and 3 fitted to the Meath data. Small values of both criteria indicate the best fitting model. DIC indicated three dimensions as the optimal model where Pritchard et al.'s criterion suggested one. Thus PCA (see Table 3) was employed as the method of selecting the optimal D.

Dimension	DIC	Pritchard et al.
1	1216797	**1209812**
2	1118221	3080834
3	**1071723**	7113250

Table 3. The proportion of variance explained by each principal component computed for configurations of the candidates in the Meath constituency, for a range of dimensions. Principal components analysis was applied to the average candidate configuration only, as the main interest of this study lies in the relative locations of the candidates.

Dimension	Variances		
	σ_1^2	σ_2^2	σ_3^2
1	1	-	-
2	0.81	0.19	-
3	0.70	0.23	0.07

and $D = 2$ appear to summarize the data well. When a three dimensional model was fitted the additional principal component only accounted for 7% of the variance of the data. Hence, we decided that the addition of an extra dimension to the model was not worthwhile because the candidate locations are concentrated around a two-dimensional plane within the three dimensional latent space.

Both the one dimensional and two dimensional configurations are analyzed to examine relationships between the electoral candidates.

5.2 One Dimensional Results

Figure 1 illustrates the one dimensional configuration of the fourteen candidates. Each candidate is represented by a two letter abbreviation of their surname and political party as detailed in Table 1. The results suggest that party politics plays an important role in the electorate's view of the candidates. The Fianna Fáil candidates (Brady, Dempsey and Wallace) are located on the far left of the single dimension with the Fine Gael candidates (Bruton, English and Farrelly) located on the far right. Fianna Fáil and Fine Gael are the two largest (and rival) Irish political parties. The other candidates lie between the two poles created by the Fianna Fáil and Fine Gael candidates but are closer to Fine Gael. Interestingly Ward, who is a Labour Party candidate, is located closest to the Fine Gael candidates — Fine Gael and Labour have a history of forming coalition governments (most recently from 1994–1997). Also of note are the narrow interval estimates for the estimated candidate positions (mean ±2 standard deviations are shown). This suggests low uncertainty in the candidate locations in one dimension.

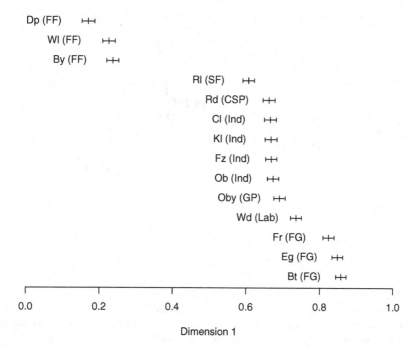

Fig. 1. The one dimensional configuration of the candidate means, averaged over a Metropolis-Hastings algorithm, and their associated uncertainty (indicated by ±2 standard deviation intervals). Each of the fourteen candidates as detailed in Table 1 are denoted by a two letter abbreviation of their surname and political party.

5.3 Two Dimensional Results

Good acceptance rates of 35% and 33% were achieved for the voter and candidate positions respectively when a two dimensional model was fitted. Figure 2 illustrates the final average position of each of the fourteen candidates in the Meath constituency. Each candidate is denoted by the abbreviations detailed in Table 1. Party politics are again suggested as the mechanism which drives this election. The first dimension (horizontal) separates candidates by their political ideals — the estimated positions suggest a clear divide between Fianna Fáil and the other parties in this dimension. The second (vertical) dimension illustrates the presence of an ideological cleavage (left to right wing) of the candidates. For example, the Christian Solidarity Party espouse right wing conservative values and their candidate Redmond (Rd) is located highest in this dimension.

The plot also includes ellipses which show approximate 95% posterior set estimates of each candidate location to represent the uncertainty in the estimated locations. The uncertainty associated with all candidate locations is low. Furthermore, there is considerable overlap between candidates from the same party.

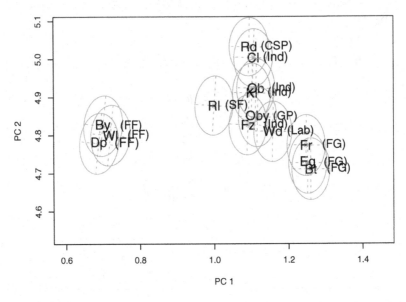

Fig. 2. The two dimensional configuration of the candidate means with their associated uncertainty. The candidate initials indicate their posterior mean positions and the ellipses are approximate 95% posterior sets which indicate the uncertainty in the candidate positions. The position of each candidate and the ellipses are estimated by 8500 Metropolis-Hastings iterations (post burn-in), thinned after every 10th iteration.

6   Conclusions and Extensions

A latent space model, incorporating a Plackett-Luce model, provides good methodology for statistically modeling PR-STV rank data. The latent space aspect of the model gives an interpretable framework for the results of model fitting and the Plackett-Luce model works well in modeling PR-STV data.

The latent configurations suggest that party politics drive general elections in Ireland. Other factors such as the level of a candidate's public profile may also be influential but some of these factors would be confounded with party membership, when it comes to an interpretation of the model's estimates.

The Plackett-Luce model is in fact a special case of Benter's model [7] which is another potential rank data model. Under Benter's model the probability of vote i is:

$$P\{\underline{x}_i | \underline{p}_i, \underline{\alpha}\} \ = \ \prod_{t=1}^{n_i} \left(\frac{p_{ic(i,t)}^{\alpha_t}}{\sum_{s=t}^{N} p_{ic(i,s)}^{\alpha_t}} \right) \qquad 1 \geq \alpha_t \geq 0$$

where α_t accounts for changing randomness at each choice level. The parameter $\underline{p}_i$ has the same interpretation as the Plackett-Luce support parameter. The α_t values 'dampen' the lower choice level probabilities to capture the randomness associated with lower choices. Gormley and Murphy [15] use a mixture of Benter's models to model data from the 1997 Irish presidential election and the

2002 Dublin West general election data. Results of that analysis also show party politics take an important role in the 2002 general election.

When defining the latent space, squared Euclidean distance was implemented as a measure of 'distance' between two members of the space. This distance worked well in the sense that the latent positions found using this distance measure are easily interpreted. Hoff [16] made use of the inner product as a latent space distance measure and such a method could be implemented in this context.

Principal components analysis selected the optimal dimension of the latent space — the method worked well but is somewhat ad-hoc. Raftery et al. [17] introduced the AICM (Akaike Information Criterion Monte (Carlo)) and BICM (Bayesian Information Criterion Monte (Carlo)) which are derived through the estimation of the harmonic mean estimator. The AICM and BICM are easily calculated from the posterior output of Monte Carlo simulations and therefore could easily be implemented as a method of estimating D. Reversible jump Metropolis-Hastings with delayed rejection is another possible but complicated method of selecting D.

In terms of the Bayesian tools used to fit the model, the random walk proposal worked well in practice but a more sophisticated proposal could be implemented. Also, a basic prior structure was used for the candidate and voter locations, yet a more structured prior on the voters could be employed — for example, a mixture of normals as was used in a social networks context by Handcock et al. [18] may provide a more suitable prior.

Acknowledgments

Both authors would like to thank Adrian Raftery and other members of the Working Group on Model-Based Clustering at the University of Washington, Seattle for important inputs into this research. We would also like to thank Vyacheslav Mikhailov and Liam Weeks from the Department of Political Science in Trinity College Dublin for providing important insights into Irish elections.

Isobel Claire Gormley was supported by a Government of Ireland Research Scholarship in Science, Engineering and Technology provided by the Irish Research Council for Science, Engineering and Technology, funded by the National Development Plan. Thomas Brendan Murphy was supported by Science Foundation of Ireland Basic Research Grant (04/BR/M0057).

The final version of this paper benefitted greatly from editor's comments.

References

[1] Hoff, P.D., Raftery, A.E., Handcock, M.S.: Latent Space Approaches to Social Network Analysis. J. Amer. Statist. Assoc. **97** (2002) 1090–1098
[2] Sinnott, R.: The electoral system. In Coakley, J., Gallagher, M., eds.: Politics in the Republic of Ireland. 3rd edn. Routledge & PSAI Press, London (1999) 99–126
[3] Coakley, J., Gallagher, M.: Politics in the Republic of Ireland. 3rd edn. Routledge in association with PSAI Press, London (1999)

[4] Sinnott, R.: Irish voters decide: Voting behaviour in elections and referendums since 1918. Manchester University Press, Manchester (1995)

[5] Mardia, K.V., Kent, J.T., Bibby, J.M.: Multivariate analysis. Academic Press, London (1979)

[6] Plackett, R.L.: The analysis of permutations. Applied Statistics **24**(2) (1975) 193–202

[7] Benter, W.: Computer-based horse race handicapping and wagering systems: A report. In Ziemba, W.T., Lo, V.S., Haush, D.B., eds.: Efficiency of Racetrack Betting Markets. Academic Press, San Diego and London (1994) 183–198

[8] Marden, J.I.: Analyzing and modeling rank data. Chapman & Hall, London (1995)

[9] Gormley, I.C., Murphy, T.B.: Analysis of Irish third-level college applications data. J. Roy. Statist. Soc. Ser. A **169**(2) (2006) 361—379

[10] Fligner, M.A., Verducci, J.S.: Multistage ranking models. J. Amer. Statist. Assoc. **83** (1988) 892–901

[11] Krzanowski, W.J.: Principles of Multivariate Analysis: A User's Perspective. Clarendon Press (1988)

[12] Airoldi, E.M., Fienberg, S.M., Joutard, C., Love, T.M.: Discovering Latent Patterns with Hierarchical Bayesian Mixed-Membership Models. Technical Report CMU-ML-06-101, School of Computer Science, Carnegie Mellon University, Pittsburgh, PA 15213, USA (2006)

[13] Spiegelhalter, D.J., Best, N.G., Carlin, B.P., van der Linde, A.: Bayesian measures of model complexity and fit. J. Roy. Statist. Soc. Ser. B **64**(4) (2002) 583–639

[14] Pritchard, J.K., Stephens, M., Donnelly, P.: Inference of Population Structure Using Multilocus Genotype Data. Genetics **155** (2000) 945–959

[15] Gormley, I.C., Murphy, T.B.: Exploring Irish Election Data: A Mixture Modelling Approach. Technical Report 05/08, Department of Statistics, Trinity College Dublin, Dublin 2, Ireland (2005)

[16] Hoff, P.D.: Bilinear Mixed-Effects Models for Dyadic Data. J. Amer. Statist. Assoc. **100** (2005) 286–295

[17] Raftery, A.E., Newton, M.A., Satagopan, J.M., Krivitsky, P.N.: Estimating the Intergrated Likelihood via Posterior Simulation Using the Harmonic Mean Identity. Technical Report 499, Department of Statistics, University of Washington, Seattle, Washington, USA. (2006)

[18] Handcock, M.S., Raftery, A.E., Tantrum, J.M.: Model-based clustering for social networks. Technical Report 482, Department of Statistics, University of Washington, Seattle, WA, USA (2005)

A Simple Model for Complex Networks with Arbitrary Degree Distribution and Clustering

Mark S. Handcock and Martina Morris*

University of Washington, Seattle WA 98195-4322, USA
handcock@stat.washington.edu
http://www.stat.washington.edu/~handcock

Abstract. We present a stochastic model for networks with arbitrary degree distributions and average clustering coefficient. Many descriptions of networks are based solely on their computed degree distribution and clustering coefficient. We propose a statistical model based on these characterizations. This model generalizes models based solely on the degree distribution and is within the curved exponential family class. We present alternative parameterizations of the model. Each parameterization of the model is interpretable and tunable. We present a simple Markov Chain Monte Carlo (MCMC) algorithm to generate networks with the specified characteristics. We provide an algorithm based on MCMC to infer the network properties from network data and develop statistical inference for the model. The model is generalizable to include mixing based on attributes and other complex social structure. An application is made to modeling a protein to protein interaction network.

1 Introduction

In this paper we consider models for relational data, and specifically networks. We have in mind social networks where the nodes represent individuals and the edges represent some form of social contact or partnership. However, the formulation is general and can be used to represent other forms of networks. We assume that the network is a realization of a stochastic process characterized by random mixing between individuals conditional on the individual activity levels (i.e., the nodal degrees) and clustering [1, 2]. One popular class are those that exhibit power-law behavior, often loosely referred to as "scale-free" distributions. We also consider models for the network degree distributions in which the variance can greatly exceed the mean.

In Section 2 we develop the general form of the model and models for the degree distribution. In Section 3 we give an simple algorithm for the generation of random networks from the model. In Section 4 we provide an algorithm for approximating the likelihood function for the model as a basis for inference. In Section 5 we apply the model to a protein-protein interaction network. Finally, in Section 6, we discuss generalizations of the model for more complex structures.

* We gratefully acknowledge the critical feedback we have received from David Hunter, Steve Goodreau and Carter T. Butts. This research supported by Grant DA012831 from NIDA and Grant HD041877 from NICHD.

E.M. Airoldi et al. (Eds.): ICML 2006 Ws, LNCS 4503, pp. 103–114, 2007.
© Springer-Verlag Berlin Heidelberg 2007

2   Models for Social Networks

2.1   Exponential Family Models

Let the random matrix X represent the adjacency matrix of an unvalued network on n individuals. We assume that the diagonal elements of X are 0 – that self-partnerships are disallowed. Suppose that $\mathcal{X}$ denotes the set of all possible networks on the given n individuals. The multivariate distribution of X can be parameterized in the form:

$$P_{\eta,\mathcal{X}}(X = x) = \frac{\exp\left[\eta \cdot Z(x)\right]}{c(\eta, \mathcal{X})} \qquad x \in \mathcal{X} \qquad (1)$$

where $\eta \in \Upsilon \subseteq \mathbb{R}^q$ is the model parameter and $Z:\mathcal{X} \to \mathbb{R}^q$ are statistics based on the adjacency matrix [3, 4]. There is an extensive literature on descriptive statistics for networks [5, 6]. These statistics are often crafted to capture features of the network (e.g., centrality, mutuality and betweenness) of primary substantive interest to the researcher. In many situations the researcher has specified a set of statistics based on substantive theoretical considerations. The above model then has the property of maximizing the entropy within the family of all distributions with given expectation of $Z(X)$ [7]. Paired with the flexibility of the choice of Z this property does provide some justification for the model (1) that will vary from application to application.

The denominator $c(\eta, \mathcal{X})$ is the normalizing function that ensures the distribution sums to one: $c(\eta, \mathcal{X}) = \sum_{y \in \mathcal{X}} \exp\left[\eta \cdot Z(y)\right]$. This factor varies with both η and the support $\mathcal{X}$ and is the primary barrier to simulation and inference under this modeling scheme.

The most commonly used class of random network models exhibit Markov dependence in the sense of [3]. For these models, dyads that do not share an individual are conditionally independent; this is an idea analogous to the nearest neighbor concept in spatial statistics. Typically a homogeneity condition is also added: all isomorphic networks have the same probability under the model. It is shown in [3] that the class of homogeneous Markov undirected networks is exactly those having the *degree parameterization*:

$$d_k(x) = \frac{\text{the proportion of nodes with}}{\text{degree exactly } k} \qquad k = 0, \ldots, n - 1$$

$$N_\Delta(x) = \frac{1}{6} \sum_{i,j,k} x_{ij} x_{jk} x_{kl},$$

where $d_k(x)$ counts the proportion of individuals with degree k and $N_\Delta(x)$ is a count of the complete triads. Throughout we consider undirected networks, although the situation for directed networks is very similar. This model can be reexpressed in the notation of model (1) by setting $Z_k(x) = d_k(x)$, $k = 1, \ldots, n - 1$, $Z_n = N_\Delta(x)$, $q = n$, $\eta \in \Upsilon = \mathbb{R}^n$. This parameterization has the advantage that it is directly interpretable in terms of concurrency of partnerships (i.e. $d_m(x)$ for $m > 0$ is the proportion of individuals with exactly m concurrent partners).

A popular variant of the statistic $N_\Delta(x)$ is the clustering coefficient defined as

$$C(x) = \frac{3N_\Delta(x)}{N_3(x)}$$

where $N_3(x)$ is the number of connected triples of nodes (i.e., 2−stars [3]). This describes the proportion of complete triads in the networks out of a total number of possible triads.

In the remainder of this paper we focus on the following novel model

$$\log\left[P_\theta(X = x)\right] = \eta(\phi) \cdot d(x) + \nu C(x) - \log c(\phi, \nu, \mathcal{X}), \qquad (2)$$

where $x \in \mathcal{X}, \theta = (\phi, \nu), \Theta \subset \mathbb{R}^n$, $d(x) = \{d_1(x), \ldots, d_{n-1}(x)\}$. The parameters ϕ and ν represent the network degree distribution and clustering, respectively. Specifically, the ratio of the probability of a given network to a network with the same degree distribution and correlation coefficient 1% less is $0.01 \times \exp(\nu)$. Alternatively, consider the conditional probability of a partnership existing given the rest of the network. If the formation of the partnership increases the correlation coefficient by $\alpha\%$ (relative to the same network without the partnership) then the log-odds of the partnership existing is $\alpha\nu\%$. The degree distribution parameters have similar interpretations: $\eta_k(\phi)$ is the ratio of the log-probability of a given network to a network with the same clustering coefficient and one less node of degree k and one more isolate. An important property of the model is the variational independence of the parameters [7].

This model is a curved exponential family if Θ is a smooth curve in $\Upsilon = \mathbb{R}^n$ [8, 9]. Any degree distribution can be specified by $n - 1$ or less independent parameters. Typically the number of parameters is small. As we shall see, this is true for the models considered below.

If $\nu = 0$ the model corresponds to random networks with arbitrary degree distributions, as considered by many researchers [10]. If $\eta_k(\phi) = \phi k, k = 1, \ldots, n-1$ the value of ϕ is interpretable as the log-probability of a given network to a network with one less partnership and the same clustering coefficient [8]. If both $\nu = 0$ and $\eta_k(\phi) = \phi k, k = 1, \ldots, n - 1$ it is the classical random network model of Rényi and Erdós[11].

The model (1) has a generative interpretation, which we illustrate with model (2). Consider a dynamic process for the network $\{X(t): t \geq 0\}$ developing according to the local rules

$$\text{logit}\left[P(X_{ij}(t) = 1|X_{ij}(t^-) = x_{ij})\right] = \eta(\phi) \cdot \left[d(x_{ij}^+) - d(x_{ij}^-)\right] + \nu\left[C(x_{ij}^+) - C(x_{ij}^-)\right]$$

where x_{ij}^+ is the network with a partnership between i and j and the rest of the network equal to x_{ij}. x_{ij}^- is similar with no partnership between i and j. Based on the theory of continuous-time Markov Chains, the equilibrium distribution is model (2). Ties are formed (or broken) based on their propensity to change the network characteristics. This also provides another interpretation of the parameters ϕ and ν and their joint effects.

An alternative parameterization that is usually more interpretable is: (ϕ, ρ) where the mapping is:

$$\rho = \mathbb{E}_{\phi,\rho}[C(X)] = \sum_{y \in \mathcal{X}} C(y) \exp[\eta(\phi) \cdot d(y) + \nu C(y)] \geq 0 \qquad (3)$$

Thus ρ is the mean clustering coefficient over networks in $\mathcal{X}$. Thus models with higher ρ have higher clustering coefficients on average. Note that models with $\rho = 0$ will not have any complete triads. The range of ρ is a subset of $[0, 1]$ and depends on the other parameters and $\mathcal{X}$.

The two parameterizations represent the same model class [9]. Translating between equivalent parameters is achieved using the MCMC algorithm given in Section 3 [9, 8].

2.2 Models for Degree Distributions

Let $P_\theta(K = k)$ be the probability mass function of K, the number of partnerships that a randomly chosen node in the network has. Based on the model (2)

$$P_\theta(K = k) = \mathbb{E}_\theta[d_k(X)] \qquad k = 0, \ldots, n - 1$$

Clearly for a given network of size n nodes, the distribution of K has finite range with upper bound $n - 1$. In some cases this distribution is approximated by an idealized distribution with infinite range. Let K^* be the degree of a node in a (possibly hypothetical) infinite population of nodes. Then K can be thought of as the degree of the node restricted to nodes in the network. In cases where this conceptualization is used we will consider the case

$$P_\theta(K = k) = P(K^* = k | K^* < n) \qquad k = 0, \ldots, n - 1,$$

While the model (2) has arbitrary degree distribution, of particular interest are the various "scale-free," preferential attachment and power-law models popular in the physics literature (see, e.g., [12]). These models assume that all networks with the same degree distribution are equally likely. We say $P(K^* = k)$ has *power-law behavior* with scaling exponent $\phi > 1$ if there exist constants c_1, c_2, and M such that $0 < c_1 \leq P(K^* = k)k^\phi \leq c_2 < \infty$ for $k > M$.

We focus on a stochastic mechanisms for the formation of the social networks that is a variation on a preferential attachment process, such as those advocated by several recent authors [13, 14]. The limiting distributions of this mechanism can be characterized by long tails.

2.3 Simple Preferential Attachment Models

A mechanism that has been suggested for the formation of power-law social networks is preferential attachment [15, 16, 2]. This and related stochastic processes have a long history in applied statistics [17, 18, 19]. Consider a population of r people in in which (1) there is a constant probability p that the $r + 1$st partnership in the population will be initiated from a randomly chosen person to a previously sexually inactive person, and (2) otherwise the probability that the $r + 1$st partnership will be to a person with exactly k partners is proportional to $kf(k|r)$, where $f(k|r)$ is the frequency of nodes with exactly k connections out of the r total links in the population. The limiting distribution of this process is known as the Waring distribution [19]. The Yule distribution discussed by [17] and used by [20] to model degree distributions is a special case of the Waring distribution with $p = (\phi_2 - 2)/(\phi_2 - 1)$.

The probability mass function (PMF) of the Waring distribution [21] is:

$$P(K^* = k) = \frac{(\phi_2 - 1)\Gamma(\phi_2 + \phi_1)}{\Gamma(\phi_1 + 1)} \cdot \frac{\Gamma(k + \phi_1)}{\Gamma(k + \phi_1 + \phi_2)}, \tag{4}$$
$$\phi_1 > -1, \phi_2 > 2,$$

where $\Gamma(\cdot)$ is the Gamma function and the mixing parameter ϕ_1 is related to p via:

$$p = \frac{\phi_2 - 2}{\phi_2 + \phi_1 - 1}. \tag{5}$$

The Waring distribution has power-law behavior with scaling exponent ϕ_2. The mean and variance of the Waring distribution are:

$$\mathbb{E}(K^*) = \frac{1}{p}, \qquad \mathbb{V}(K^*) = \frac{(1 - p)(\phi_2 - 1)}{p^2(\phi_2 - 3)}, \qquad \phi_2 > 3.$$

Thus, the expected value of the Waring distribution is simply the inverse of the probability of forming a partnership to an individual lacking existing partnerships. Both the Waring and the Yule distributions have been re-discovered, apparently without awareness of their historical antecedents, by [22] and [23] respectively in the context of modeling growth of the Internet.

3 Generating Random Networks with Specified Structure

Markov Chain Monte Carlo (MCMC) algorithms for generating from the model (1) have a long history and been well studied (see [24] for a review). The basic idea is to generate a Markov chain whose stationary distribution is given by equation (1). The simplest Markov chain proceeds by choosing (by some method, either stochastic or deterministic) a dyad (i, j) and then deciding whether to set $X_{ij} = 1$ or $X_{ij} = 0$ at the next step of the chain. One way to do this is using Gibbs sampling, whereby the new value of X_{ij} is sampled from the conditional distribution of X_{ij} conditional on the rest of the network. Denote "the rest of the network" by X_{ij}^c. Then $X_{ij}|X_{ij}^c = x_{ij}^c$ has a Bernoulli distribution, with odds given by

$$\frac{P(X_{ij} = 1|X_{ij}^c = x_{ij}^c)}{P(X_{ij} = 0|X_{ij}^c = x_{ij}^c)} = \exp\{\eta \cdot \Delta(Z(x))_{ij}\},$$

where $\Delta(Z(x))_{ij}$ denotes the difference between $Z(x)$ when x_{ij} is set to 1 and $Z(x)$ when x_{ij} is set to 0. A simple variant to the Gibbs sampler (which is an instance of a Metropolis-Hastings algorithm) is a pure Metropolis algorithm in which the proposal is always to change the value of x_{ij}. This proposal is accepted with probability $\min\{1, \pi\}$, where

$$
\pi = \frac{P(X_{ij} = 1 - x_{ij} | X_{ij}^c = x_{ij}^c)}{P(X_{ij} = x_{ij} | X_{ij}^c = x_{ij}^c)} \tag{6}
$$

$$
= \begin{cases} \exp\{\eta \cdot \Delta(Z(x))_{ij}\} & \text{if } x_{ij} = 0; \\ \exp\{-\eta \cdot \Delta(Z(x))_{ij}\} & \text{if } x_{ij} = 1. \end{cases}
$$

The vector $\Delta(Z(x))_{ij}$ used by these MCMC schemes is often much easier to calculate directly than as the difference of two separate values of $Z(x)$. For instance, if one of the components of the $Z(x)$ vector is the total number of partnerships in the network, then the corresponding component of $\Delta(Z(x))_{ij}$ is always equal to 1.

The Metropolis scheme is usually preferred over the Gibbs scheme because it results in a greater probability of changing the value of x_{ij}, a property thought to produce better-mixing chains. However, it is well known that these simple MCMC schemes often fail for various reasons to produce well-mixed chains [25, 26, 27]. More sophisticated MCMC schemes have been developed and are a topic of ongoing research [8].

A variant of this algorithm proceeds in two steps:

1. Generate $d_k \overset{\text{i.i.d.}}{\sim} P_\theta(K = k)$, $k = 0, 1, \ldots, n - 1$.
2. Generate a random network conditional on this degree distribution:

$$
P_\nu(X = x | d_k(X) = d_k) = \frac{\exp[\nu C(x)]}{c(\nu, d_k, \mathcal{X})} \quad x \in \mathcal{X}(d_k)
$$

where $\mathcal{X}(d_k) = \{x \in \mathcal{X} : d_k(x) = d_k\}$.

The first generates individual degrees from an arbitrary distribution, and the second generates networks condition on those degrees. Note that the structure of the exponential family in (1) ensure that the samples are from the correct distribution [7]. The first step can be simulated easily as we know $P_\theta(K = k)$. Note that not all degree sequences will be consistent with a network of size n. For example, sequences with an odd total number of partnerships are not realizable. However we can construct a compatible sequence $\{d_k\}_{k=0}^{n-1}$ via a simple rejection algorithm. The second step is also straightforward: we can conditionally simulate values using a MCMC holding the degree distribution fixed by using a Metropolis proposal consistent with this restriction. It is convenient for this algorithm to have a starting network with the given degree distribution. This network is easy to construct by a finite algorithm (as it need not be be a draw from a random distribution) or using sequential importance sampling. An important property of this the second step is the independence of the distribution from ϕ. It is a simple parameter distribution depending only on ν [7].

As an application of this algorithm, consider a network model for $n = 50$ nodes. We choose a degree distribution which is Yule with scaling exponent $\phi_2 = 3$. This

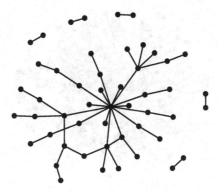

Fig. 1. An example network generated from model (2) with $n = 50$ and degree distribution draw from the Yule model (equation 4) with scaling exponent $\phi_2 = 3$. The random network is drawn from the model with mean clustering coefficient $\rho = 3\%$. The network has clustering coefficient $C(x) = 2\%$.

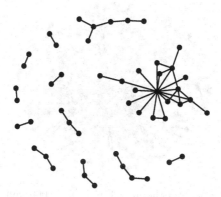

Fig. 2. An example network generated from model (2) with $n = 50$ and degree distribution draw from the Yule model (equation 4) with scaling exponent $\phi_2 = 3$. The random network is drawn from the model with mean clustering coefficient $\rho = 15\%$. The network has clustering coefficient $C(x) = 18\%$.

corresponds to a "scale-free" degree model. If $\nu = 0$ the network is random with the given degree distribution. This corresponds to a mean clustering coefficient $\rho = 3\%$. A realization of this model is given in Fig. 1. The clustering coefficient for this network is 2%. Fig. 2 is a realization from the model with mean clustering coefficient $\rho = 15\%$ (corresponding to a clustering parameter of $\nu = 0.46$.) The centralization of the clustering is apparent relative to the network in Fig. 1.

As an second application we generate a network model for $n = 1000$ nodes with the same degree distribution ($\phi_2 = 3$). A realization of this model is given in Fig. 3. The clustering coefficient for this network is 2%. Fig. 4 is a realization from the model with mean clustering coefficient chosen to be $\rho = 15\%$ (corresponding

Fig. 3. A random network from the model with $n = 1000$ and the same Yule degree distribution with $\phi_2 = 3$.. The largest component is visualized. The network is drawn from the model with mean clustering coefficient $\rho = 3\%$. The realized network has clustering coefficient $C(x) = 1\%$.

Fig. 4. A random network from the model with $n = 1000$ and the same Yule degree distribution with $\phi_2 = 3$.. The largest component is visualized. The network is drawn from the model with mean clustering coefficient $\rho = 15\%$. The realized network has clustering coefficient $C(x) = 14\%$.

to a clustering parameter of $\nu = 27$.) The elongated nature of the resulting network is apparent as is the centralization of the clustering.

4   Statistical Inference for Network Models

As we have specified the full joint distribution of the network through (1), we choose to conduct inference within the likelihood framework [28, 24]. For economy of notation, in this section, we use ϕ to represent either η in (1) or the curved exponential family form (ϕ, ν) in (2). Differentiating the loglikelihood function:

$$\ell(\phi; x) \equiv \log\left[P_\eta(X = x)\right] = \eta(\phi) \cdot Z(x) - \log\left[c(\phi, \mathcal{X})\right] \qquad (7)$$

shows that the maximum likelihood estimate $\hat{\phi}$ satisfies

$$\nabla \ell(\hat{\phi}) = \nabla \eta(\hat{\phi}) \cdot \left[Z(x_{\text{obs}}) - \mathrm{E}_{\eta(\hat{\phi})} Z(X) \right], \tag{8}$$

where $\nabla \eta(\phi)$ is the $p \times q$ matrix of partial derivatives of η with respect to ϕ and $Z(x_{\text{obs}})$ is the observed network statistic. We may search for a solution to equation (8) using an iterative technique such as Newton-Raphson; however, the exponential family form of the model makes the Fisher information matrix

$$I(\phi) = \nabla \eta(\phi) \cdot \left[\mathrm{Cov}_{\eta(\phi)} Z(X) \right] \nabla \eta(\phi) \tag{9}$$

easier to calculate than the Hessian matrix of second derivatives required for Newton-Raphson. For more about equations (8) and (9), see [8] The method of Fisher scoring is an iterative method analogous to Newton-Raphson except that the negative Fisher information is used in place of the Hessian matrix.

Direct calculation of the log-likelihood by enumerating $\mathcal{X}$ is infeasible for all but the smallest networks. As an alternative, we approximate the likelihood equations (8) by replacing the expectations by (weighted) averages over a sample of networks generated from a known distribution. This procedure is described in [24]. To generate the sample we use the MCMC algorithm of Section 3.

5   Application to a Protein-Protein Interaction Network

As an application of these methods, we fit the model to a biological network of protein-protein interactions found in cells. By interact is meant that two amino acid chains were experimentally identified to bind to each other. The network is for *E. Coli* and is drawn from the "Database of Interacting Proteins (DIP)" [29]. The DIP database lists protein pairs that are known to interact with each other. The dataset we use is `Ecoli20050403`. We have chosen *E. Coli* as it is well studied and this will minimize the number of false-negative interactions (that is, two proteins that interact but are not in the database). For simplicity we focus on proteins that interact with themselves and have at least one other interaction. We do not represent the self-interactions as part of the network. This results in a network in Figure 5 with 108 proteins and 94 interactions.

We consider the model (2) with a clustering coefficient term and the degree distribution model by a preferential attachment process (the Yule distribution with scaling exponent ϕ). We choose the Yule as it represents the simple version of preferential attachment that is common in the literature. The estimates are given in Table 1. They are derived using the algorithm in Section 4.

The estimate of the preferential attachment scaling decay rate of about three suggests that the network is close to the so-called "scale-free" range (that is, $\phi \leq 3$). We note that the standard errors are based on the curvature of the estimated log-likelihood and approximations to the sampling distribution based on asymptotic arguments require non-standard justifications. In this case the standard approximation to the sampling distribution can be shown to be valid using a parametric bootstrap. The standard error of the scaling rate indicates

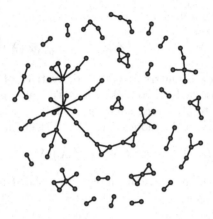

Fig. 5. A protein - protein interaction network for *E. Coli*. The nodes represent proteins and the partnerships indicate that the two proteins are known to interact with each other.

Table 1. MCMC maximum likelihood parameter estimates for the protein-protein interaction network

Parameter	est.	s.e.
Scaling decay rate (ϕ)	3.034	0.3108
Correlation Coefficient (ν)	1.176	0.1457

some uncertainty in the determination of the rate. However the parameter of the correlation coefficient is very positive. This indicates strong clustering (given the degree sequence) and hence so-called "small world" behavior in the network. Thus, this model provides a statistical valid means to test for small-world characteristics of a network using the statistics commonly used to characterize small-world networks.

Finally, we can test if the network is generated by this preferential attachment model. If preferential attachment among proteins generated this network then the parameter ν of the clustering coefficient will be zero. However we see that the estimate is positive. We can test this more rigorously by comparing the log-likelihood values for the maximum likelihood fit in Table 1 to the model where ν is constrained to be zero. The change in the log-likelihood is 52.3, so that the change in deviance is 104.6. This indicates that deviation from the preferential attachment model is statistically significant, as can be verified by a parametric bootstrap of the change in deviance.

6   Discussion

We have presented a simple stochastic model for random networks that has arbitrary degree distribution and average clustering coefficient. The clustering component of the model is directly interpretable via the clustering coefficient of the realizations from the model. The model places positive probability over the set of possible networks. Conditional on the degree sequence, the clustering coefficient covers the full range of values possible. The distribution over this range is tuned as a monotone function of the clustering parameter.

The model form (1) is very general, and can incorporate general social structure [3, 30, 9, 8]. For example, in disease epidemiology, the two-sex random network epidemic model is a commonly used to represent the contact structure of pathogens transmitted by intimate contact [31]. This model is the model (2) with $\rho = 0$ and $\mathcal{X}$ is restricted to heterosexual networks. However, this model contains a major weakness which ultimately limits its utility. Specifically, it assumes random mixing conditional on degree. The model (2) is a simple extension of that allows tunable correlation coefficient. More generally, (1) can be used to include nodal attributes and other structural characteristics. Such models have proven to be valuable in epidemiology [32, 33].

References

[1] Newman, M.E.J.: Spread of epidemic disease on networks. Physical Review E **66**(1) (2002) art. no.–016128

[2] Dezsó, Z., Barabási, A.L.: Halting viruses in scale-free networks. Physical Review E **65** (2002) art. no. 055103

[3] Frank, O., Strauss, D.: Markov graphs. Journal of the American Statistical Association **81**(395) (1986) 832–842

[4] Handcock, M.S.: Degeneracy and inference for social network models. In: Paper presented at the Sunbelt XXII International Social Network Conference in New Orleans, LA. (2002)

[5] Wasserman, S.S., Faust, K.: Social Network Analysis: Methods and Applications. Cambridge: Cambridge University Press (1994)

[6] Borgatti, S.P., Everett, M.G., Freeman, L.C.: Ucinet 5.0 for Windows. Natick: Analytic Technologies (1999)

[7] Barndorff-Nielsen, O.E.: Information and Exponential Families in Statistical Theory. Wiley, New York (1978)

[8] Hunter, D.R., Handcock, M.S.: Inference in curved exponential family models for networks. Journal of Computational and Graphical Statistics (2006) to appear

[9] Handcock, M.S.: Assessing degeneracy in statistical models of social networks. Working paper #39, Center for Statistics and the Social Sciences, University of Washington (2003)

[10] Newman, M.E.J., Strogatz, S.H., Watts, D.J.: Random graphs with arbitrary degree distributions and their applications. Physical Review E **64** (2001) 026118

[11] Bollobas, B.: Random Graphs. Academic Press, NY, USA (1985)

[12] Newman, M.E.J.: The structure and function of complex networks. SIAM Review **45**(2) (2003) 167–256

[13] Barabási, A.L., Albert, R.: Emergence of scaling in random networks. Science **286**(5439) (1999) 509–512

[14] Pastor-Satorras, R., Vespignani, A.: Epidemic dynamics and endemic states in complex networks. Physical Review E **63**(6) (2001) art. no.–066117

[15] Albert, R., Barabási, A.L.: Topology of evolving networks: Local events and universality. Physical Review Letters **85**(24) (2000) 5234–5237

[16] Liljeros, F., Edling, C.R., Amaral, L.A.N., Stanley, H.E., Åberg, Y.: The web of human sexual contacts. Nature **411**(6840) (2001) 907–908

[17] Simon, H.: On a class of skew distribution functions. Biometrika **42**(3/4) (1955) 435–440

[18] Kendall, M.: Natural law in the social sciences: Presidential address. Journal of the Royal Statistical Society Series A-Statistics in Society **124**(1) (1961) 1–16

[19] Irwin, J.: The place of mathematics in medical and biological statistics. Journal of the Royal Statistical Society Series A (General) **126**(1) (1963) 1–45

[20] Jones, J., Handcock, M.S.: An assessment of preferential attachment as a mechanism for human sexual network formation. Proceedings of the Royal Society of London, B **270** (2003) 1123–1128

[21] Johnson, N., Kotz, S., Kemp, A.: Univariate discrete distributions. 2nd edn. Wiley series in probability and mathematical statistics. Wiley, New York (1992)

[22] Levene, M., Fenner, T., Loizou, G., Wheeldon, R.: A stochastic model for the evolution of the web. Computer Networks **39**(3) (2002) 277–287

[23] Dorogovtsev, S.N., Mendes, J.F.F., Samukhin, A.N.: Structure of growing networks with preferential linking. Physical Review Letters **85**(21) (2000) 4633–4636

[24] Geyer, C.J., Thompson, E.A.: Constrained monte carlo maximum likelihood calculations (with discussion). Journal of the Royal Statistical Society, Series B **54** (1992) 657–699

[25] Snijders, T.A.B.: Markov chain monte carlo estimation of exponential random graph models. Journal of Social Structure **3**(2) (2002)

[26] Handcock, M.S.: Progress in statistical modeling of drug user and sexual networks. Manuscript, Center for Statistics and the Social Sciences, University of Washington (2000)

[27] Snijders, T.A.B., Pattison, P., Robins, G.L., Handcock, M.S.: New specifications for exponential random graph models. Sociological Methodology **36** (2006) to appear

[28] Besag, J.: Statistical analysis of non-lattice data. The Statistician **24** (1975) 179–95

[29] Xenarios, I., Salwinski, L., Duan, X.J., Higney P., Kim, S., Eisenberg, D.: Dip, the database of interacting proteins: a research tool for studying cellular networks of protein interactions. Nucleic Acids Res **28** (2002) 289–291

[30] Strauss, D., Ikeda, M.: Pseudolikelihood estimation for social networks. Journal of the American Statistical Association **85** (1990) 204–212

[31] Newman, M.E.J.: Assortative mixing in networks. Physical Review Letters **89**(20) (2002) art no. – 208701

[32] Morris, M.: Local rules and global properties: Modeling the emergence of network structure. In Breiger, R., Carley, K., Pattison, P., eds.: Dynamic Social Network Modeling and Analysis. Committee on Human Factors, Board on Behavioral, Cognitive, and Sensory Sciences. National Academy Press: Washington, DC (2003) 174–186

[33] Handcock, M.S.: Statistical models for social networks: Inference and degeneracy. In Breiger, R., Carley, K., Pattison, P., eds.: Dynamic Social Network Modeling and Analysis. Committee on Human Factors, Board on Behavioral, Cognitive, and Sensory Sciences. National Academy Press: Washington, DC (2003) 229–240

Discrete Temporal Models of Social Networks

Steve Hanneke and Eric P. Xing

Machine Learning Department
Carnegie Mellon University
Pittsburgh, PA 15213 USA
{shanneke,epxing}@cs.cmu.edu

Abstract. We propose a family of statistical models for social network evolution over time, which represents an extension of Exponential Random Graph Models (ERGMs). Many of the methods for ERGMs are readily adapted for these models, including MCMC maximum likelihood estimation algorithms. We discuss models of this type and give examples, as well as a demonstration of their use for hypothesis testing and classification.

1   Introduction

The field of social network analysis is concerned with populations of *actors*, interconnected by a set of *relations* (e.g., friendship, communication, etc.). These relationships can be concisely described by directed graphs, with one vertex for each actor and an edge for each relation between a pair of actors. This network representation of a population can provide insight into organizational structures, social behavior patterns, emergence of global structure from local dynamics, and a variety of other social phenomena.

There has been increasing demand for flexible statistical models of social networks, for the purposes of scientific exploration and as a basis for practical analysis and data mining tools. The subject of modeling a static social network has been investigated in some depth. In particular, there is a rich (and growing) body of literature on the *Exponential Random Graph Models* (ERGM) [1, 2, 3, 4]. Specifically, if N is some representation of a social network, and $\mathcal{N}$ is the set of all possible networks in this representation, then the probability distribution function for any ERGM can be written in the following general form.

$$\mathcal{P}(N) = \frac{1}{Z(\boldsymbol{\theta})} \exp\left\{\boldsymbol{\theta}'\mathbf{u}(N)\right\}.$$

Here, $\boldsymbol{\theta} \in \mathbb{R}^k$, and $\mathbf{u} : \mathcal{N} \to \mathbb{R}^k$. $Z(\boldsymbol{\theta})$ is a normalization constant, which is typically intractable to compute. The $\mathbf{u}$ function represents the sufficient statistics for the model, and, in a graphical modeling interpretation, can be regarded as a vector of clique potentials. The representation for N can vary widely, possibly including multiple relation types, valued or binary relations, symmetric or asymmetric relations, and actor and relation attributes. The most widely studied models of this form are for single-relation social networks, in which case N is

E.M. Airoldi et al. (Eds.): ICML 2006 Ws, LNCS 4503, pp. 115–125, 2007.
© Springer-Verlag Berlin Heidelberg 2007

generally taken to be the weight matrix A for the network (sometimes referred to as a *sociomatrix*), where A_{ij} is the strength of directed relation between the ith actor and j^{th} actor.

Often one is interested in modeling the evolution of a network over multiple sequential observations. For example, one may wish to model the evolution of coauthorship networks in a specific community from year to year, trends in the evolution of the World Wide Web, or a process by which simple local relationship dynamics give rise to global structure. In the following sections, we propose a model family that is capable of modeling network evolution, while maintaining the flexibility of ERGMs. Furthermore, these models build upon ERGMs, so that existing methods developed for ERGMs over the past two decades are readily adapted to apply to the temporal models as well.

2   Discrete Temporal Models

We begin with the simplest case of the proposed models, before turning to the fully general derivation. One way to simplify a statistical model for social networks is to make a Markov assumption on the network from one time step to the next. Specifically, if A^t is the weight matrix representation of a single-relation social network at time t, then we can assume A^t is independent of $A^1, \ldots, A^{t-2}$ given A^{t-1}. Put another way, a sequence of network observations $A^1, \ldots, A^t$ has the property that

$$\mathcal{P}(A^2, A^3, \ldots, A^t | A^1) = \mathcal{P}(A^t | A^{t-1}) \mathcal{P}(A^{t-1} | A^{t-2}) \cdots \mathcal{P}(A^2 | A^1).$$

With this assumption in mind, we can now set about deciding what the form of the conditional PDF $\mathcal{P}(A^t | A^{t-1})$ should be. Given our Markov assumption, one natural way to generalize ERGMs for evolving networks is to assume $A^t | A^{t-1}$ admits an ERGM representation. That is, we can specify a function $\boldsymbol{\Psi} : \mathbb{R}_{n \times n} \times \mathbb{R}_{n \times n} \to \mathbb{R}^k$ and parameter vector $\boldsymbol{\theta} \in \mathbb{R}^k$, such that the conditional PDF has the following form.

$$\mathcal{P}(A^t | A^{t-1}, \boldsymbol{\theta}) = \frac{1}{Z(\boldsymbol{\theta}, A^{t-1})} \exp \left\{ \boldsymbol{\theta}' \boldsymbol{\Psi}(A^t, A^{t-1}) \right\} \tag{1}$$

2.1   An Example

To illustrate the expressivity of this framework, we present the following simple example model. For simplicity, assume the weight matrix is binary (i.e., an adjacency matrix). Define the following statistics, which represent *density*, *stability*, *reciprocity*, and *transitivity*, respectively.

$$\Psi_D(A^t, A^{t-1}) = \frac{1}{(n-1)} \sum_{ij} A_{ij}^t$$

$$\Psi_S(A^t, A^{t-1}) = \frac{1}{(n-1)} \sum_{ij} \left[A_{ij}^t A_{ij}^{t-1} + (1 - A_{ij}^t)(1 - A_{ij}^{t-1}) \right]$$

$$\Psi_R(A^t, A^{t-1}) = n \left[\sum_{ij} A^t_{ji} A^{t-1}_{ij} \right] \Big/ \left[\sum_{ij} A^{t-1}_{ij} \right]$$

$$\Psi_T(A^t, A^{t-1}) = n \left[\sum_{ijk} A^t_{ik} A^{t-1}_{ij} A^{t-1}_{jk} \right] \Big/ \left[\sum_{ijk} A^{t-1}_{ij} A^{t-1}_{jk} \right]$$

The statistics are each scaled to a constant range (in this case $[0, n]$) to enhance interpretability of the model parameters. The conditional probability mass function (1) is governed by four parameters: θ_D controls the *density*, or the number of ties in the network as a whole; θ_S controls the *stability*, or the tendency of a link that does (or does not) exist at time $t - 1$ to continue existing (or not existing) at time t; θ_R controls the *reciprocity*, or the tendency of a link from i to j to result in a link from j to i at the next time step; and θ_T controls the *transitivity*, or the tendency of a tie from i to j and from j to k to result in a tie from i to k at the next time step. The transition probability for this temporal network model can then be written as follows.

$$P(A^t|A^{t-1}, \boldsymbol{\theta}) = \frac{1}{Z(\boldsymbol{\theta}, A^{t-1})} \exp \left\{ \sum_{j \in \{D, S, R, T\}} \theta_j \Psi_j(A^t, A^{t-1}) \right\}$$

2.2   General Models

We can generalize the form of (1) by replacing $A^1, A^2, \ldots, A^T$ with general networks $N^1, N^2, \ldots, N^T \in \mathcal{N}$, which may include multiple relations, actor attributes, etc. Furthermore, we generalize the Markov assumption to allow any K-order dependencies, so that the previous discussion was for $K = 1$. In this case, the function $\boldsymbol{\Psi}$ is also generalized by $\boldsymbol{\Psi} : \mathcal{N}^{K+1} \to \mathbb{R}^k$. The fully general model can therefore be written as

$$P(N^{K+1}, N^{K+2}, \ldots, N^T|N^1, \ldots, N^K, \boldsymbol{\theta}) = \prod_{t=K+1}^{T} P(N^t|N^{t-K}, \ldots, N^{t-1}, \boldsymbol{\theta}),$$

where

$$P(N^t|N^{t-K}, \ldots, N^{t-1}, \boldsymbol{\theta}) = \frac{1}{Z(\boldsymbol{\theta}, N^{t-K}, \ldots, N^{t-1})} \exp\{\boldsymbol{\theta}' \boldsymbol{\Psi}(N^t, N^{t-1}, \ldots, N^{t-K})\}.$$

Note that specifying the joint distribution requires one to specify a distribution over the first K networks. This can generally be accomplished fairly naturally using an ERGM for N^1 and exponential family conditional distributions for $N^i|N^1 \ldots N^{i-1}$ for $i \in \{2, \ldots, K\}$. For simplicity of presentation, we avoid these details in subsequent sections by assuming the distribution over these initial K networks is functionally independent of the parameter $\boldsymbol{\theta}$.

3 Estimation

The estimation task for models of the above form is to use the sequence of observed networks, $N^1, N^2, \ldots, N^T$, to find an estimator $\hat{\boldsymbol{\theta}}$ that is close to the actual parameter values $\boldsymbol{\theta}$ in some sensible metric. As with ERGMs, the intractability of the normalizing constant Z often makes explicit solution of maximum likelihood estimation difficult. However, general techniques for MCMC sampling to enable approximate maximum likelihood estimation for ERGMs have been studied in some depth and have proven successful for a variety of models [3]. By a slight modification of these algorithms, we can apply the same general techniques as follows.

Let

$$\mathcal{L}(\boldsymbol{\theta}; N^1, N^2, \ldots, N^T) = \log \mathcal{P}(N^{K+1}, N^{K+2}, \ldots, N^T | N^1, \ldots, N^K, \boldsymbol{\theta}), \quad (2)$$

$$\mathbf{M}(t, \boldsymbol{\theta}) = \mathbb{E}_{\boldsymbol{\theta}} \left[\Psi(\underline{\mathbf{N}}^t, N^{t-1}, \ldots, N^{t-K}) | N^{t-1}, \ldots, N^{t-K} \right],$$

$$\mathbf{C}(t, \boldsymbol{\theta}) = \mathbb{E}_{\boldsymbol{\theta}} \left[\Psi(\underline{\mathbf{N}}^t, N^{t-1}, \ldots, N^{t-K}) \Psi(\underline{\mathbf{N}}^t, N^{t-1}, \ldots, N^{t-K})' | N^{t-1}, \ldots, N^{t-K} \right].$$

where expectations are taken over the random variable $\underline{\mathbf{N}}^t$, the network at time t. Note that

$$\nabla \mathcal{L}(\boldsymbol{\theta}; N^1, \ldots, N^T) = \sum_{t=K+1}^{T} \left(\Psi(N^t, N^{t-1}, \ldots, N^{t-K}) - M(t, \boldsymbol{\theta}) \right)$$

and

$$\nabla^2 \mathcal{L}(\boldsymbol{\theta}; N^1, \ldots, N^T) = \sum_{t=K+1}^{T} \left(M(t, \boldsymbol{\theta}) M(t, \boldsymbol{\theta})' - C(t, \boldsymbol{\theta}) \right).$$

The expectations can be approximated by Gibbs sampling from the conditional distributions [3], so that we can perform an unconstrained optimization procedure akin to Newton's method: approximate the expectations, update parameter values in the direction that increases (2), repeat until convergence. A related algorithm is described by [5] for general exponential families, and variations are given by [3] that are tailored for ERG models. The following is a simple version of such an estimation algorithm.

---

1. Randomly initialize $\theta^{(1)}$
2. For $i = 1$ up until convergence
3.   For $t = K+1, K+2, \ldots, T$
4.   Sample $\hat{N}_{(i)}^{t,1}, \ldots, \hat{N}_{(i)}^{t,B} \sim \mathcal{P}(\underline{\mathbf{N}}^t | N^{t-K}, \ldots, N^{t-1}, \theta^{(i)})$
5. $\hat{\mu}_{(i)}^t = \frac{1}{B} \sum_{b=1}^{B} \Psi(\hat{N}_{(i)}^{t,b}, N^{t-1}, \ldots, N^{t-K})$
6. $\hat{C}_{(i)}^t = \frac{1}{B} \sum_{b=1}^{B} \Psi(\hat{N}_{(i)}^{t,b}, N^{t-1}, \ldots, N^{t-K}) \Psi(\hat{N}_{(i)}^{t,b}, N^{t-1}, \ldots, N^{t-K})'$
7. $\hat{H}_{(i)} = \sum_{t=K+1}^{T} [\hat{\mu}_{(i)}^t \hat{\mu}_{(i)}^{t\prime} - \hat{C}_{(i)}^t]$
8. $\theta^{(i+1)} \leftarrow \theta^{(i)} - \hat{H}_{(i)}^{-1} \sum_{t=K+1}^{T} \left[\Psi(N^t, N^{t-1}, \ldots, N^{t-K}) - \hat{\mu}_{(i)}^t \right]$

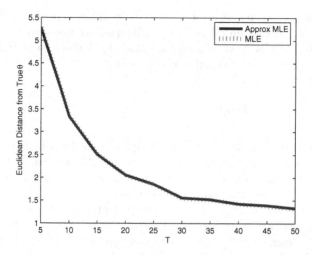

Fig. 1. Convergence of estimation algorithm on simulated data, measured in Euclidean distance of the estimated values from the true parameter values. The approximate MLE from the sampling-based algorithm is almost identical to the MLE obtained by direct optimization.

The choice of B can affect the convergence of this algorithm. Generally, larger B values will give more accurate updates, and thus fewer iterations needed until convergence. However, in the early stages of the algorithm, precise updates might not be necessary if the likelihood function is sufficiently smooth, so that a B that grows larger only when more precision is needed may be appropriate. If computational resources are limited, it is possible (though less certain) that the algorithm might still converge even for small B values (see [6] for an alternative approach to sampling-based MLE, which seems to remain effective for small B values).

To examine the convergence rate empirically, we display in Figure 1 the convergence of this algorithm on data generated from the example model given in Section 2.1. The simulated data is generated by sampling from the example model with randomly generated θ, and the loss is plotted in terms of Euclidean distance of the estimator from the true parameters. To generate the initial N^1 network, we sample from the pmf $\frac{1}{Z(\theta)} \exp\{\theta' \Psi(N^1, N^1)\}$. The number of actors n is 100. The parameters are initialized uniformly in the range $[0, 10)$, except for θ_D, which is initialized to $-5\theta_S - 5\theta_R - 5\theta_T$. This tends to generate networks with reasonable densities. The results in Figure 1 represent averages over 10 random initial configurations of the parameters and data. In the estimation algorithm used, $B = 100$, but increases to 1000 when the Euclidean distance between parameter estimates from the previous two iterations is less than 1. Convergence is defined as the Euclidean distance between $\theta^{(i+1)}$ and $\theta^{(i)}$ being within 0.1. Since this particular model is simple enough for exact calculation of the likelihood and derivatives thereof (see below), we also compare against Newton's method with exact updates (rather than sampling-based). We can use this to determine how much of the loss is due to the approximations being performed, and how much of it is intrinsic to the estimation problem. The parameters returned by the sampling-based approximation are usually almost identical to the MLE obtained by Newton's method, and this behavior manifests itself in Figure 1 by the losses being visually indistinguishable.

4   Hypothesis Testing

As an example of how models of this type might be used in practice, we present a simple hypothesis testing application. Here we see the generality of this framework pay off, as we can use models of this type to represent a broad range of scientific hypotheses. The general approach to hypothesis testing in this framework is first to write down potential functions representing transitions one expects to be of some significance in a given population, next to write down potential functions representing the usual "background" processes (to serve as a null hypothesis), and third to plug these potentials into the model, calculate a test statistic, and compute a p-value.

The data involved in this example come from the United States 108th Senate, having $n = 100$ actors. Every time a proposal is made in the Senate, be it a bill, amendment, resolution, etc., a single Senator serves as the proposal's *sponsor* and there may possibly be several *cosponsors*. Given records of all proposals voted on in the full Senate, we create a sliding window of 100 consecutive proposals. For a particular placement of the window, we define a binary directed relation existing between two Senators if and only if one of them is a sponsor and the other a cosponsor for the same proposal within that window (where the direction is toward the sponsor). The data is then taken as evenly spaced snapshots of this sliding window, A^1 being the adjacency matrix for the first 100 proposals, A^2 for proposal 31 through 130, and so on shifting the window by 30 proposals each time. In total, there are 14 observed networks in this series, corresponding to the first 490 proposals addressed in the 108th Senate.

In this study, we propose to test the hypothesis that intraparty reciprocity is inherently stronger than interparty reciprocity. To formalize this, we use a model similar to the example given previously. The main difference is the addition of party membership indicator variables. Let $P_{ij} = 1$ if the ith and j^{th} actors are in the same political party, and 0 otherwise, and let $\bar{P}_{ij} = 1 - P_{ij}$. Define the following potential functions, representing *stability, intraparty density, interparty density,*[1] *overall reciprocity, intraparty reciprocity,* and *interparty reciprocity.*

$$\Psi_S(A^t, A^{t-1}) = \frac{1}{(n-1)} \sum_{ij} \left[A_{ij}^t A_{ij}^{t-1} + (1-A_{ij}^t)(1-A_{ij}^{t-1}) \right]$$

$$\Psi_{WD}(A^t, A^{t-1}) = \frac{1}{(n-1)} \sum_{ij} A_{ij}^t P_{ij}$$

---

[1] We split density to intra- and inter-party terms so as to factor out the effects on reciprocity of having higher intraparty density.

$$\Psi_{BD}(A^t, A^{t-1}) = \frac{1}{(n-1)} \sum_{ij} A_{ij}^t \bar{P}_{ij}$$

$$\Psi_R(A^t, A^{t-1}) = n \left[\sum_{ij} A_{ji}^t A_{ij}^{t-1} \right] \bigg/ \left[\sum_{ij} A_{ij}^{t-1} \right]$$

$$\Psi_{WR}(A^t, A^{t-1}) = n \left[\sum_{ij} A_{ji}^t A_{ij}^{t-1} P_{ij} \right] \bigg/ \left[\sum_{ij} A_{ij}^{t-1} P_{ij} \right]$$

$$\Psi_{BR}(A^t, A^{t-1}) = n \left[\sum_{ij} A_{ji}^t A_{ij}^{t-1} \bar{P}_{ij} \right] \bigg/ \left[\sum_{ij} A_{ij}^{t-1} \bar{P}_{ij} \right]$$

The null hypothesis supposes that the reciprocity observed in this data is the result of an overall tendency toward reciprocity amongst the Senators, regardless of party. The alternative hypothesis supposes that there is a stronger tendency toward reciprocity among Senators within the same party than among Senators from different parties. Formally, the transition probability for the null hypothesis can be written as

$$P_0(A^t | A^{t-1}, \boldsymbol{\theta}^{(0)}) = \frac{1}{Z_0(\boldsymbol{\theta}^{(0)}, A^{t-1})} \exp \left\{ \sum_{j \in \{S, WD, BD, R\}} \theta_j^{(0)} \Psi_j(A^t, A^{t-1}) \right\},$$

while the transition probability for the alternative hypothesis can be written as

$$P_1(A^t | A^{t-1}, \boldsymbol{\theta}^{(1)}) = \frac{1}{Z_1(\boldsymbol{\theta}^{(1)}, A^{t-1})} \exp \left\{ \sum_{j \in \{S, WD, BD, WR, BR\}} \theta_j^{(1)} \Psi_j(A^t, A^{t-1}) \right\}.$$

For our test statistic, we use the likelihood ratio. To compute this, we compute the maximum likelihood estimators for each of these models, and take the ratio of the likelihoods. For the null hypothesis, the MLE is

$$(\hat{\theta}_S^{(0)} = 336.2, \hat{\theta}_{WD}^{(0)} = -58.0, \hat{\theta}_{BD}^{(0)} = -95.0, \hat{\theta}_R^{(0)} = 4.7)$$

with likelihood value of $e^{-9094.46}$. For the alternative hypothesis, the MLE is

$$(\hat{\theta}_S^{(1)} = 336.0, \hat{\theta}_{WD}^{(1)} = -58.8, \hat{\theta}_{BD}^{(1)} = -94.3, \hat{\theta}_{WR}^{(1)} = 4.2, \hat{\theta}_{BR}^{(1)} = 0.03)$$

with likelihood value of $e^{-9088.96}$. The likelihood ratio statistic (null likelihood over alternative likelihood) is therefore about 0.0041. Because the null hypothesis is composite, determining the p-value of this result is a bit more tricky, since we must determine the probability of observing a likelihood ratio at least this extreme under the null hypothesis for the parameter values $\boldsymbol{\theta}^{(0)}$ that *maximize* this probability. That is,

$$\text{p-value} = \sup_{\boldsymbol{\theta}^{(0)}} P_0 \left\{ \frac{\sup_{\hat{\boldsymbol{\theta}}^{(0)}} P_0(A^1, \ldots, A^{14} | \hat{\boldsymbol{\theta}}^{(0)})}{\sup_{\hat{\boldsymbol{\theta}}^{(1)}} P_1(A^1, \ldots, A^{14} | \hat{\boldsymbol{\theta}}^{(1)})} \leq 0.0041 \,\bigg|\, \boldsymbol{\theta}^{(0)} \right\}.$$

In general this seems not to be tractable to analytic solution, so we employ a genetic algorithm to perform the unconstrained optimization, and approximate the probability for each parameter vector by sampling. That is, for each parameter vector $\boldsymbol{\theta}^{(0)}$ (for the null hypothesis) in the GA's population on each iteration, we sample a large set of sequences from the joint distribution. For each sequence, we compute the MLE under the null hypothesis and the MLE under the alternative hypothesis, and then calculate the likelihood ratio and compare it to the observed ratio. We calculate the empirical frequency with which the likelihood ratio is at most 0.0041 in the set of sampled sequences for each vector $\boldsymbol{\theta}^{(0)}$, and use this as the objective function value in the genetic algorithm. Mutations consist of adding Gaussian noise (with variance decreasing on each iteration), and recombination is performed as usual. Full details of the algorithm are omitted for brevity (see [7] for an introduction to GAs). The resulting approximate p-value we obtain by this optimization procedure is 0.024.

This model is nice in that we can compute the likelihoods and derivatives thereof analytically. In fact, it is representative of an interesting subfamily of models, in which the distributions of edges at time t are independent of each other given the network at time $t-1$. In models of this form, we can compute likelihoods and perform Newton-Raphson optimization directly, without the need of sampling-based approximations. However, in general this might not be the case. For situations in which one cannot tractably compute the likelihoods, an alternative possibility is to use bounds on the likelihoods. Specifically, one can obtain an upper bound on the likelihood ratio statistic by dividing an upper bound on the null likelihood by a lower bound on the alternative likelihood. When computing the p-value, one can use a lower bound on the ratio by dividing a lower bound on the null likelihood by an upper bound on the alternative likelihood. See [8, 9] for examples of how such bounds on the likelihood can be tractably attained, even for intractable models.

In practice, the problem of formulating an appropriate model to encode one's hypothesis is ill-posed. One general approach which seems intuitively appealing is to write down the types of motifs or patterns one expects to find in the data, and then specify various other patterns which one believes those motifs could likely transition to (or would likely *not* transition to) under the alternative hypothesis. For example, perhaps one believes that densely connected regions of the network will tend to become more dense and clique-like over time, so that one might want to write down a potential representing the transition of, say, k-cliques to more densely connected structures.

5   Classification

One can additionally consider using these temporal models for classification. Specifically, consider a transductive learning problem in which each actor has a static class label, but the learning algorithm is only allowed to observe the labels of some random subset of the population. The question is then how to use the known label information, in conjunction with observations of the network evolving over time, to accurately infer the labels of the remaining actors whose labels are unknown.

As an example of this type of application, consider the alternative hypothesis model from the previous section (model 1), in which each Senator has a class label (party affiliation). We can slightly modify the model so that the party labels are no longer constant, but random variables drawn independently from a known multinomial distribution. Assume we know the party affiliations of a randomly chosen 50 Senators. This leaves 50 Senators with unknown affiliations. If we knew the parameters θ, we could predict these 50 labels by sampling from the posterior distribution and taking the mode for each label. However, since *both* the parameters *and* the 50 labels are unknown, this is not possible. Instead, we can perform Expectation Maximization to *jointly* infer the maximum likelihood estimator $\hat{\theta}$ for θ *and* the posterior mode given $\hat{\theta}$.

Specifically, let us assume the two class labels are *Democrat* and *Republican*, and we model these labels as independent Bernoulli(0.5) random variables. The distribution on the network sequence given that all 100 labels are fully observed is the same as given in the previous section. Since one can compute likelihoods in this model, sampling from the posterior distribution of labels given the network sequence is straightforward using Gibbs sampling. We can therefore employ a combination of MCEM and Generalized EM algorithms (call it MCGEM) [10] with this model to infer the party labels as follows. In each iteration of the algorithm, we sample from the posterior distribution of the unknown class labels under the current parameter estimates given the observed networks and known labels, approximate the expectation of the gradient and Hessian of the log likelihood using the samples, and then perform a single Newton-Raphson update using these approximations.

We run this algorithm on the 108th Senate data from the previous section. We randomly select 50 Senators whose labels are observable, and take the remaining Senators as having unknown labels. As mentioned above, we assume all Senators are either Democrat or Republican; Senator Jeffords, the only independent Senator, is considered a Democrat in this model. We run the MCGEM algorithm described above to infer the maximum likelihood estimator $\hat{\theta}$ for θ, and then sample from the posterior distribution over the 50 unknown labels under that maximum likelihood distribution, and take the sample mode for each label to make a prediction.

The predictions of this algorithm are correct on 70% of the 50 Senators with unknown labels. Additionally, it is interesting to note that the parameter values the algorithm outputs ($\hat{\theta}_S = 336.0, \hat{\theta}_{WD} = -59.7, \hat{\theta}_{BD} = -96.0, \hat{\theta}_{WR} = 3.8, \hat{\theta}_{BR} = 0.28$) are very close (Euclidean distance 2.0) to the maximum likelihood estimator obtained in the previous section (where all class labels were known). Compare the above accuracy score with a baseline predictor that always predicts Democrat, which would get 52% correct for this train/test split, indicating that this statistical model of network evolution provides at least a somewhat reasonable learning bias. However, there is clearly room for improvement in the model specification, and it is not clear whether modeling the

Table 1. Summary of classification results

Method	Accuracy
Baseline	52%
Temporal Model	70%
SGT	90%

evolution of the graph is actually of any benefit for this particular example. For example, after collapsing this sequence of networks into a single weighted graph with edge weights equal to the sum of edge weights over all graphs in the sequence, running Thorsten Joachims' Spectral Graph Transducer algorithm [11] gives a 90% prediction accuracy on the Senators with unknown labels. These results are summarized in Table 1. Further investigation is needed into what types of problems can benefit from explicitly modeling the network evolution, and what types of models are most appropriate for basing a learning bias on.

6   Open Problems and Future Work

If we think of this type of model as describing a process giving rise to the networks one observes in reality, then one can think of a single network observation as a snapshot of this Markov chain at that time point. Traditionally one would model a network at a single time point using an ERGM. It therefore seems natural to investigate the formal relation between these Markov chain models and ERGMs. Specifically, any Markov chain of the form described here has a stationary distribution which can be characterized by an ERGM. Can one give a general analytic derivation of this stationary ERGM for any Markov chain of the form described here? To our knowledge, this remains an open problem. One can also ask the reverse question of whether, given any ERGM, one can describe an interesting set of Markov chains having it as a stationary distribution. Answering this would not only be of theoretical interest, but would potentially also lead to practical techniques for sampling from an ERGM distribution by formulating a more tractable Markov chain giving rise to it. Indeed, one can ask these same questions about general Markov chains (not just networks) having transition probabilities in an exponential family, the stationary distributions of which can be described by exponential families.

Moving forward, we hope to move beyond these ERG-inspired models toward models that incorporate latent variables, which may also evolve over time with the network. For example, it may often be the case that the phenomena represented in data can most easily be described by imagining the existence of unobserved groups or factions, which form, dissolve, merge and split as time progresses. The flexibility of the ERG models and the above temporal extensions allows a social scientist to "plug in" his or her knowledge into the formulation of the model, while still providing general-purpose estimation algorithms to find the right trade-offs between competing and complementary factors in the model. We would like to retain this flexibility in formulating a general family of models that include evolving latent variables in the representation, so that the researcher can "plug in" his or her hypotheses about latent group dynamics, evolution of unobservable actor attributes, or a range of other possible phenomena into the model representation. At the same time, we would like to preserve the ability to provide a "black box" inference algorithm to determine the parameter and variable values of interest to the researcher, as can be done with ERG models and their temporal extensions.

Acknowledgments

We thank Stephen Fienberg and all participants in the statistical network modeling discussion group at Carnegie Mellon for helpful comments and feedback. We also thank the anonymous reviewers for insightful suggestions.

References

[1] Anderson, C.J., Wasserman, S., Crouch, B.: A p* primer: Logit models for social networks. Social Networks **21** (1999) 37–66

[2] Robins, G.L., Pattison, P.E.: Interdependencies and social processes: Generalized dependence structures. Cambridge University Press (2004)

[3] Snijders, T.A.B.: Markov Chain Monte Carlo estimation of exponential random graph models. Journal of Social Structure **3**(2) (2002)

[4] Frank, O., Strauss, D.: Markov graphs. Journal of the American Statistical Association **81** (1986)

[5] Geyer, C., Thompson, E.: Constrained Monte Carlo maximum likelihood for dependent data. Journal of the Royal Statistical Society, B **54**(3) (1992) 657–699

[6] Carreira-Perpignán, M., Hinton, G.E.: On contrastive divergence learning. In: Artificial Intelligence and Statistics. (2005)

[7] Mitchell, M.: An Introduction to Genetic Algorithms. MIT Press (1996)

[8] Opper, M., Saad, D., eds.: Advanced Mean Field Methods: Theory and Practice. MIT Press (2001)

[9] Wainwright, M., Jaakkola, T., Willsky, A.: A new class of upper bounds on the log partition function. IEEE Trans. on Information Theory **51**(7) (2005)

[10] McLachlan, G., Krishnan, T.: The EM algorithm and Extensions. Wiley (1997)

[11] Joachims, T.: Transductive learning via spectral graph partitioning. In: Proceedings of the International Conference on Machine Learning. (2003)

Approximate Kalman Filters for Embedding Author-Word Co-occurrence Data over Time

Purnamrita Sarkar[1], Sajid M. Siddiqi[2], and Geoffrey J. Gordon[1]

[1] Machine Learning Department,
Carnegie Mellon University, Pittsburgh, PA 15213
{psarkar,ggordon}@cs.cmu.edu
[2] Robotics Institute,
Carnegie Mellon University, Pittsburgh, PA 15213
siddiqi@cs.cmu.edu

Abstract. We address the problem of embedding entities into Euclidean space over time based on co-occurrence data. We extend the CODE model of [1] to a dynamic setting. This leads to a non-standard factored state space model with real-valued hidden parent nodes and discrete observation nodes. We investigate the use of variational approximations applied to the observation model that allow us to formulate the entire dynamic model as a Kalman filter. Applying this model to temporal co-occurrence data yields posterior distributions of entity coordinates in Euclidean space that are updated over time. Initial results on per-year co-occurrences of authors and words in the NIPS corpus and on synthetic data, including videos of dynamic embeddings, seem to indicate that the model results in embeddings of co-occurrence data that are meaningful both temporally and contextually.

1 Introduction

Embedding discrete entities into Euclidean space is an important area of research for obtaining interpretable representations of relationships between objects. This is very useful for visualization, clustering and exploratory data analysis. Recent work [1] proposes a novel technique for embedding heterogeneous entities such as author-names and paper keywords into a single Euclidean space based on their co-occurrence counts. When applied to the NIPS corpus, the resulting clusters of keywords and authors reflect real-life relationships between different research areas and researchers in those respective areas. However, it would be interesting to see how these relationships evolve over time, an aspect which these techniques do not address. Recent work has examined the dynamic behavior of social networks [2], but only with homogeneous entities, and with point estimates of the embedding coordinates. The problem we are interested in differs in two ways: first, embedding time-series co-occurrence data from two kinds of entities (essentially weighted link data from a bipartite graph) in a dynamic model could be useful for temporal data visualization, link prediction and group detection in such networks. Examples of such bipartite data are author-word co-occurrences

E.M. Airoldi et al. (Eds.): ICML 2006 Ws, LNCS 4503, pp. 126–139, 2007.

in conference proceedings over time, actor-director collaborations throughout their careers, and so on. Second, modelling a *distribution* over the coordinates of these embeddings instead of point estimates (as in [2]) would tell us about the correlation and uncertainty in the entities' coordinates. In this paper, we explore one possible approach to achieve both these goals.

The layout of the rest of this paper is as follows. We discuss some related work, in particular the model of [1] which we utilize. We then extend this model to the dynamic case, describing how our dynamic model can be used for posterior estimation using a Kalman filter after some approximations. The resulting model keeps track of the belief state over all author and word coordinates in the latent space based on the approximated co-occurrence observation model and a zero-mean Gaussian transition model. We give derivations and intuition for the operation of this dynamic model, as well as results on the NIPS corpus of author-word co-occurrence data and on synthetic data.

2   Related Work

The problem of embedding discrete entities into euclidean space is well-studied. Principal Components Analysis (PCA) is a standard technique based on eigen-decomposition of the counts matrix [3]. Multi-Dimensional Scaling (MDS) [4] is another technique. However, these techniques are not suitable for temporal data if one wishes to enforce smoothness constraints on embeddings over time.

[5] introduced a model similar to MDS in which entities are associated with locations in p-dimensional space, and links are more likely if the entities are close in latent space. However their work does not take the sequential aspect of the data into account. Also, the distribution over latent positions are obtained by sampling, which becomes intractable for large networks. Their work also assumes binary link data.

The most closely related work is the CODE model of [1], which gives a technique for embedding heterogenous entities (such as authors and keywords) based on co-occurence data for the static case. We briefly introduce their model here, and our notation is similar to theirs.

The basic model of CODE is a conditional model $p(w|a)$, where w denotes the words and a denotes the authors. Let ϕ_i and ψ_j denote the hidden variables representing the coordinates of author a_i and word w_j in the latent space respectively. By $\Phi_t(A)$, $\Psi_t(W)$ we represent the states related to all author and word positions at timestep t. The conditional probability of seeing word w_j given an author a_i is related (inversely) to the distance $d_{ij} = |\phi_i - \psi_j|$ of author i and word j in the latent space, as well as the marginal counts of each individual entity, $\bar{p}(a_i)$ and $\bar{p}(w_j)$. For latent coordinates in a d dimensional space,

$$p(w_j|a_i) = \frac{\bar{p}(w_j)}{Z(a_i)} e^{-|\phi_i - \psi_j|^2}$$
$$Z(a_i) = \sum_{w_j} \bar{p}(w_j) e^{-|\phi_i - \psi_j|^2} \qquad (1)$$
$$|\phi_i - \psi_j|^2 = \sum_{k=1}^{d} (\phi_i^k - \psi_j^k)^2$$

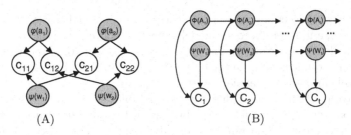

Fig. 1. Shaded nodes indicate hidden random variables. (A) The graphical model relating author/keyword positions to co-occurrence counts at a single timestep. (B) The corresponding factored state-space model for temporal inference.

The hidden coordinates $\Phi_t(A)$, $\Psi_t(W)$ are learned by maximizing the likelihood objective function using conjugate gradient or other such techniques.

3   The Single-Timestep Model

The original conditional model was chosen by considering $\frac{p(w|a)}{\bar{p}(w)}$ to be inversely proportional to the exponentiated squared distance between the latent embeddings $\phi(a)$ and $\psi(w)$. Similarly, our model of the joint is motivated by considering the initial ratio to be $\frac{p(w,a)}{\bar{p}(w)\bar{p}(a)}$ instead, and deriving the resultant $p(w,a)$. The reason for dividing by the empirical marginals is to normalize the joint by the overall frequencies of the individual entities in the joint. This represents the single timestep graphical model shown in Figure 1(A). The resultant $p(w,a)$ is as follows:

$$p(a_i, w_j | \phi_i, \psi_j) = \frac{1}{Z}\bar{p}(a_i)\bar{p}(w_j)e^{-|\phi_i - \psi_j|^2}$$
$$Z = \sum_{a_i}\sum_{w_j} \bar{p}(a_i)\bar{p}(w_j)e^{-|\phi_i - \psi_j|^2} \tag{2}$$

4   Dynamic Embedding of Co-occurrence Data Through Time

We consider the unknown coordinates of authors and words to be hidden variables in a latent space. Our goal is now to estimate these continuous hidden variables given discrete co-occurrence observations. As shown above, we model the joint posterior probability of author and word coordinates (given the observations) based on the distances between those coordinates. To make the problem tractable, we aim to derive a Gaussian distribution that is somehow close to our observation model, which would allow us to use Kalman Filters, which are described below. The natural approach which we follow is to minimize the KL-divergence of a Gaussian distribution (as an approximation to the observation model) and the normalized likelihood of our model. However, this turns out to be difficult since the KL-divergence has no closed-form solution, mainly due to the non-standard $\log(Z)$ term (where Z is defined in equation (2). We investigate two methods for making this expression tractable and obtaining a

Gaussian that approximates the observation model. We will see how the approximated model, together with a Gaussian transition model for the coordinates, can be formulated as a standard dynamic model.

4.1   The State-Space Model

For our state-space model in the dynamic setting, we choose a factored state space model as shown in Figure 1(B), similar to a factorial HMM [6] or switching state space model [7]. It is a natural choice over the full joint model because we consider the hidden coordinates of authors and words to be decoupled Markov chains conditionally coupled given their co-occurrence. This model closely resembles the factorial HMM model yet is distinct because of the hidden variables being real-valued. Exact filtering and smoothing are very difficult in this model because the prior belief state is not conjugate to the discrete observation density for typical belief distribution choices like the Normal distribution. Instead, we would like to approximate this exact model in order to formulate it as a Kalman Filter.

4.2   Kalman Filters

A Kalman filter [8] is a linear chain graphical model with a backbone of hidden real-valued states emitting a real-valued observation at every timestep. Both the observation and transition models are assumed to be Gaussian. It is commonly used in tracking the states of complex systems or locations of moving objects such as robots or missiles. Filtering and smoothing are tractable in this model because of the conjugacy of the Gaussian distribution to itself, which enables the belief state to remain Normally distributed at each timestep after the three standard steps of *conditioning* (factoring in a new observation to the current belief state), *prediction* (propogating the belief through the transition model) and *rollup* (marginalizing to obtain the new belief state). These steps are described in more detail below.

4.3   Kalman Filter Formulation for Dynamic Embedding

In a standard Kalman Filter, all three steps mentioned above have closed form solutions, i.e.:

$$\text{Conditioning: } P(\Phi_t, \Psi_t | C_{1:t-1}, C_t = c_t)$$
$$\propto P(C_t = c_t | \Phi_t, \Psi_t) P(\Phi_t, \Psi_t | C_{1:t-1})$$

$$(3)$$

$$\text{Prediction and Rollup: } P(\Phi_{t+1}, \Psi_{t+1} | C_{1:t})$$
$$= \int_{\Phi_t} \int_{\Psi_t} P(\Phi_{t+1}, \Psi_{t+1} | \Phi_t, \Psi_t) P(\Phi_t, \Psi_t | C_{1:t}) \partial \Phi_t \partial \Psi_t$$

These are the Kalman filter updates in our model. Lets see what happens for our model in the conditioning step. The observation model is:

$$\log p(C_t | \Phi_t, \Psi_t)$$
$$= -\sum_{a_i} \sum_{w_j} \bar{p}(a_i, w_j) |\phi_{t,i} - \psi_{t,j}|^2 - \log Z$$

$$(4)$$

However, this is not a Gaussian kernel, so we do not have a closed form update equation available. Now we look at approximations to project this family of density functions to a Gaussian, in order to overcome this problem.

4.4   Approximate Conditioning Step

A simple approach: Jensen's Inequality. One natural approach is to apply Jensen's inequality to approximate the difficult portion of the likelihood (i.e. the $\log Z$ term), which happens to be concave. However as we shall see, this approximation causes us to lose much of the information encoded in the normalization constant, and will not be used in our final model. The log normalizing function of our joint model is

$$log Z = log(\sum_{a_i} \sum_{w_j} \bar{p}(a_i)\bar{p}(w_j)e^{-||\phi_{t,i}-\psi_{t,j}||^2}) \tag{5}$$

Using Jensen's inequality,

$$log Z \geq -\sum_{a_i} \sum_{w_j} \bar{p}(a_i)\bar{p}(w_j)||\phi_{t,i} - \psi_{t,j}||^2 \tag{6}$$

This gives us a lower bound on the KL divergence between an approximate Gaussian distribution p and our distribution q. We denote $p(a_i)$ by p_i and $p(w_j)$ by p_j. We also denote by χ the random variables $< \Phi, \Psi >$. Maximizing the KL divergence (details in the Appendix) gives us the parameters for the closest Gaussian approximation to our observation model with mean zero and covariance Σ given by the following equation.

$$\Sigma^{-1} = 2\hat{\Lambda} \tag{7}$$

Where $\hat{\Lambda}$ is defined as follows:

$$\hat{\Lambda}_{ij} = \begin{cases} \sum_j \tilde{c}_{ij}I_{2\times 2} & j = i,\, 1 \leq i \leq 2A - 1 \\ \sum_i \tilde{c}_{ij}I_{2\times 2} & i = j,\, 2A + 1 \leq j \leq 2(A + W) - 1 \\ -2\tilde{c}_{ij}I_{2\times 2} & i \neq j,\, 1 \leq i \leq 2A - 1, \\ & 2A + 1 \leq j \leq 2(A + W) - 1 \\ 0_{2\times 2} & \text{otherwise} \end{cases} \tag{8}$$

In the above equation $\tilde{c}_{ij} = \bar{p}_{ij} - \bar{p}_i\bar{p}_j$. Note that there is no correlation between the x and y coordinates in this model. It is clear that the numerator of our observation model doesn't give rise to any such correlation.

However the log-normalization constant gives rise to such correlation, which is clear from figure 2. Unfortunately this approximation removes the correlations between the x, y coordinates as we can see from equation 8. Having uncorrelated x and y coordinates implies that higher-dimensional embeddings are not beneficial, and that we may as well be embedding to a line. In practice, this model often leaves us with such an embedding even when the space is two-dimensional,

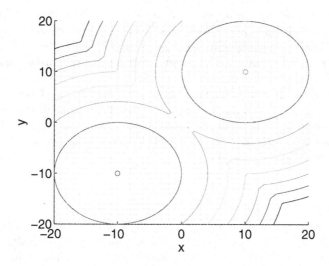

Fig. 2. A plot of the log normalizing constant $log(e^{(-(x-a)^2-(y-b)^2)}+e^{(-(x-c)^2-(y-d)^2)})$ for two given coordinates a, b and c, d. Two things are apparent: the correlation of x and y coordinates , and the presence of multiple optima in this function. We desire an approximation that preserves the $x - y$ correlation.

since we are optimizing over the two dimensions independently. Also the mean of the observation model is zero. Also this method is effectively minimizing a lower bound on the KL divergence, which is not necessarily beneficial. We therefore look for a better model.

A more sophisticated approach: Taylor approximation of a variational upper bound. Now we try and come up with a model which preserves the correlations between the axes. We look at a variational upper bound on the log normalizing constant [9].

$$\log Z \leq \lambda \sum_{ij} \overline{p}_i \overline{p}_j e^{-(\phi_i - \psi_j)^T (\phi_i - \psi_j)} - 1 - \log \lambda$$

Minimizing this upper bound effectively minimizes an upper bound on the KL-divergence. However, direct minimization of this bound is difficult because of the term inside the expectation, and because the expression is not convex. Instead, we take a second order Taylor approximation of the $e^{-(\phi_i - \psi_j)^T (\phi_i - \psi_j)}$ values around ξ_i, ξ_j. A Taylor approximation of a function $g(x)$ is given by,

$$g(x) = g(0) + x^T [\frac{\partial g}{\partial x_1}, \frac{\partial g}{\partial x_2}]_{\xi_i, \xi_j} + \frac{1}{2} x^T H(\xi_i, \xi_j) x$$

Where $H(\xi_i, \xi_j)$ is the Hessian of the function evaluated at ξ_i, ξ_j.

Now we have a Gaussian approximation to our observation model, which has canonical parameters Λ, η. These parameters, as derived in the appendix, are functions of the Jacobian and Hessian matrix of the taylor approximation, evaluated at ξ_i, ξ_j. We shall describe how we choose these parameters later in this section.

In (3), we multiply two Gaussians i.e. prior $p(\Phi_t, \Psi_t | C_{1:t-1})$ with canonical parameters $(\eta_{t|t-1}, \Lambda_{t|t-1})$ and the approximate observation distribution with η, Λ. The notation $\eta_{t|t-1}$ denotes the value of a parameter at time t conditioned on observations from timesteps $1 \ldots t-1$. The resulting Gaussian $p(\Phi_t, \Psi_t | C_{1:t})$ is distributed with $\eta_{t|t}, \Lambda_{t|t}$, where

$$\eta_{t|t} = \eta_{t|t-1} + \eta$$
$$\Lambda_{t|t} = \Lambda_{t|t-1} + \Lambda$$

We compute the moment parameters $\mu_{t|t}, \Sigma_{t|t}$ from the canonical parameters. And we get the $\eta_{t|t-1}, \Lambda_{t|t-1}$ from the previous time-step of the Kalman Filter.

When applying the Taylor expansion, we set the ξ values to the $\mu_{t|t-1}$ learnt from the previous timestep. We found this to be most effective, and this also makes sense since given the former time-steps' data we are most likely to be around the conditional means predicted from the former time-steps. Because of the noncon-vex structure of the log-normalizer, which is due to the presence of saddle points (Figure 2), the resulting Λ can become non-positive definite and have negative eigenvalues. To project to the closest possible positive definite matrix, we set the negative eigenvalues to zero (plus a small positive constant). Together these ap-proximations succeed in giving us a tractable expression while not losing the highly informative inter-coordinate interactions (e.g. x-y correlation in two dimensions) that the simple Jensen's inequality approach would discard.

4.5 Prediction and Rollup Step

Our transition model is very simple, just a zero-mean symmetric increase in uncertainty:

$$(\Phi_{t+1}, \Psi_{t+1}) = (\Phi_t, \Psi_t) + N(0, \Sigma_{transition})$$

Here $\Sigma_{transition}$ is a diagonal noise term denoting the spread of uncertainty along both axes, which must be fixed beforehand. The prediction and rollup steps give the following result:

$$(\Phi_{t+1}, \Psi_{t+1}) \sim N(\mu_{t+1|t}, \Sigma_{t+1|t})$$

where $\mu_{t+1|t} = \mu_{t|t}$ and $\Sigma_{t+1|t} = \Sigma_{t|t} + \Sigma_{transition}$.

4.6 Computational Issues

Note that we model all author-word interactions with a *single* large Kalman filter, where the authors and words relate through the covariance matrix. This introduces complexity issues since the size of the covariance matrix is propor-tional to the number of authors and words. However some sparseness properties of the covariance matrix can be exploited for faster computation.

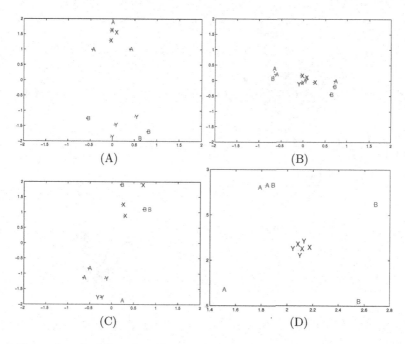

Fig. 3. Dynamic embedding of synthetic data vs. static embedding. A, B are two groups of authors and X, Y are two groups of words. The 140-timestep data smoothly varies from strong A-X and B-Y links to strong A-Y and B-X links. The entities are initialized randomly (not shown). A. $t = 20$, strong A-X and B-Y links. B. $t = 70$, Intermediate configuration, noisy uniform links. C. Strong A-Y and B-X links. D. A static embedding of the aggregate co-occurrence matrix, which is effectively a noisy uniform matrix, resulting in entities mixing with each other.

5   Experiments

We divide the results section in three parts. We present some snapshots from our algorithm on embeddings of a synthetic datasets with pre-specified dynamic structure. We then present snapshots and closeups of embeddings of author-word co-occurrence data from the NIPS corpus over thirteen years. We also show how the distance in our embedding between author-word pairs in the corpus evolve over time. In all cases, $\Sigma_{transition}$ is currently set heuristically to give a smoothly varying embedding that is still responsive to new data. We finish our experimental section with a comparison with PCA [3], a well-studied static embedding technique.

5.1   Modeling Trends over Time

We wish to inspect the performance of dynamic embedding in cases where the underlying model is known. To do this, we generate noisy co-occurrence matrices of 3 words and 3 authors over 140 timesteps. The matrices have some amount

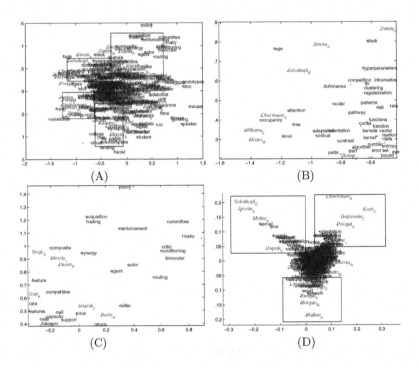

Fig. 4. (A). $t = 13$ Dynamic embedding of NIPS data (final timestep, 1999). (B),(C). Close-ups of (roughly) the top two rectangles in (A). The first Both contain authors and keyword groups that are interrelated (e.g. (B) contains entities related to kernels, (C) contains reinforcement-learning-related terms and authots. (D). PCA embedding of aggregate counts matrix of NIPS data, that averages out any sequential patterns.

of random sparseness in every timestep, to be more realistic. We divide the authors in two groups, namely A, B and the words in two groups X, Y. We vary the co-occurrences between these groups smoothly such that in the first 20 steps, authors A have high co-occurrence counts with X, and B with Y, whereas the A-Y and B-X counts are very low. After $t = 20$, this pattern starts becoming less sharp, blending to a completely uniform matrix with noise at $t = 70$. From then until $t = 120$, the authors and words "switch" i.e. A-Y and B-X counts begin to dominate. From $t = 120$ to 140, the data continues to reflect strong A-Y and B-X co-occurrences. A movie with this and other dynamic embeddings is available at http://www.cs.cmu.edu/~psarkar/icml06/. Figure 3(A,B,C) shows three snapshots from a dynamic embedding of this data sequence, which clearly reflect the underlying dynamic structure at different timesteps. In contrast, Figure 3(D) shows a static embedding of the aggregate summed counts matrix, which happens to be approximately uniform and thus not indicative of any interesting structure in the data.

5.2   The NIPS Corpus

In this section we shall look at word-author co-occurrence data over thirteen years from the NIPS proceedings of 1986-1999. We implemented the dynamic Kalman filter models on a subset of the NIPS dataset. The NIPS data corpus[1] contains co-occurrence count data for 13, 649 words and 2, 037 authors appearing together in papers from 1986 to 1999. We partitioned this data into yearly raw count matrices using additional information in the dataset, and picked a set of well-known authors and meaningful keywords. The experiments shown here are carried out on small subsets of authors and words in order to get easily interpretable 2-D plots for this paper, however the algorithm scales well to larger sets.

Qualitative Analysis. The resulting embedding has some very interesting properties. The words on different parts of it define different areas of machine learning. We also find the corresponding authors in those areas. For example in figure 4(A) we have presented the embedding of 40 authors and 428 words. These are the overall most popular authors, and the words they tend to use.

We can divide the area in the figure in four clear areas, within the rectangles. The top right region magnified in Figure 4(C) has words like reinforcement, agent, actor, policy which clearly are words from the field of reinforcement learning. We also have authors such as Singh, Dayan and Barto in the same area. Dayan is known to have worked on acquisition and trading which are also words in this region. However the very neighboring region on the left belongs to words like kernel, regularization, error and bound. We see some overlap with that region via the entities support and Vapnik. Also one of the other two interesting regions consists of authors Jordan, Hinton, Gharamani Zemel, Tresp. The lowest rectangular region is filled with words and authors like image, segmentation, motion, movement. Notably we find that author Viola is placed very close to these words and words like document, retrieval,facial. Also we have author Murray co-placed with words voltage, circuit, chip, analog, synapse. These are strongly supported by the co-occurrence data and anecdotal evidence.

Quantitative Analysis. A single embedding does not tell us whether our algorithm models dynamic structure. To investigate this aspect, in Figure 5 we plot the average distance per timestep between three word-author pairs of interest, along with the empirical probability of that pair per timestep, to see whether the distances correlate to the probabilities. As we can see in the bottom panels of Figures 5, (Jordan,variational) and (Smola,kernel) have high empirical probabilities in the later timesteps, corresponding to drops in the distance between these entities' coordinates. In contrast, (Waibel,speech) co-occurs mostly in the first half of the data set, and so we see the distance between the author-word embeddings shrinking initially then gradually increasing over time.

---

[1] http://www.cs.toronto.edu/ ∼ roweis/data.html

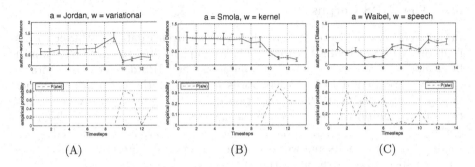

Fig. 5. Average distance between author-word pairs over time (above), along with corresponding empirical probabilities (below). A. Jordan and variational. B. Smola and kernel C. Waibel and speech. The graphs on the bottom reflect empirical $\bar{p}(author \mid word)$ from the NIPS data which varies inversely over time with the average author-word distance in the embedding shown in the top row, demonstrating the responsiveness of the embeddings to the underlying data.

5.3 Comparison with PCA

An embedding of the aggregate data with PCA is shown in Figure 4(D). The embedding reflects relationships in the overall data very well, as seen in the three rectangles highlighted. For example, one of them has entities like `Scholkopf`, `Smola`, `kernel` and `pca`, and the others also have consistent sets of authors and the keywords they are known to use. However the data fails to capture dynamic trends in the data that our model successfully reflects. For example, `Waibel` and `speech` do not co-occur at all in the latter timesteps of the dataset, as is clear from the lower panel of Figure 5(C). However, since the aggregate counts matrix embedded by static PCA averages out all sequential structure, `Waibel` and `speech` are still relatively close in the PCA embedding.

6   Conclusion and Future Work

We have proposed and demonstrated a model for Euclidean embedding of co-occurrence data over time by formulating the problem as a factored state space model, and used an approximation to yield a tractable Kalman filter formulation. The resulting model gives us an estimate of the posterior distribution over the coordinates of the entities in latent space. The previous work we are extending addresses this problem only for the single-timestep case, giving only point estimates for the coordinates. Experimental results show that our model yields interpretable visual results and reflects dynamic trends in the data. For future work we will implement smoothing in the dynamic model to see if it offers improved results over filtering. We will also obtain quantitative results for the model on problems such as link prediction in social networks and classification in word-document embedding.

Acknowledgements

We warmly thank Carlos Guestrin for his guidance. This work was funded in part by DARPA's CS2P program under grant number HR0011-006-1-0023. The opinions and conclusions expressed are the authors'.

References

1. Globerson, A., Chechik, G., Pereira, F., Tishby, N.: Euclidean embedding of co-occurrence data. In: Proc. Eighteenth Annual Conf. on Neural Info. Proc. Systems (NIPS). (2004)
2. Sarkar, P., Moore, A.: Dynamic social network analysis using latent space models. In: Proc. Nineteenth Annual Conf. on Neural Info. Proc. Systems (NIPS). (2005)
3. Berry, M., Dumais, S., Letsche, T.: Computational methods for intelligent information access. In: Proceedings of Supercomputing. (1995)
4. Breiger, R.L., Boorman, S.A., Arabie, P.: An algorithm for clustering relational data with applications to social network analysis and comparison with multidimensional scaling. J. of Math. Psych. **12** (1975) 328–383
5. Raftery, A.E., Handcock, M.S., Hoff, P.D.: Latent space approaches to social network analysis. J. Amer. Stat. Assoc. **15** (2002) 460
6. Ghahramani, Z., Jordan, M.I.: Factorial hidden Markov models. In Touretzky, D.S., Mozer, M.C., Hasselmo, M.E., eds.: Proc. Conf. Advances in Neural Information Processing Systems, NIPS. Volume 8., MIT Press (1995) 472–478
7. Ghahramani, Z., Hinton, G.E.: Switching state-space models. Technical report, 6 King's College Road, Toronto M5S 3H5, Canada (1998)
8. Kalman, R.: A new approach to linear filtering and prediction problems. (1960)
9. Jordan, M.I., Ghahramani, Z., Jaakkola, T.S., Saul, L.K.: An Introduction to Variational Methods for Graphical Methods. Machine Learning (1998)

Appendix

In this section we give a detailed description of the derivations.

Derivation of Section 4.4

We compute the KL projection of our observation model (p) to the closest Gaussian family (q).

$$
\begin{aligned}
D(p,q) &= \int p \ln p - \int p \ln q \\
&= -H(p) + \int (\sum_{ij} \bar{p}_{ij} (\phi_i - \psi_j)^T (\phi_i - \psi_j)) dp + E_p(\ln Z) \\
&= -(A+W) - \frac{\ln((2\pi)^{2(A+W)}|\Sigma|)}{2} \\
&\quad + E_p(\sum_{ij} \bar{p}_{ij} (\phi_i - \psi_j)^T (\phi_i - \psi_j)) + E_p(\ln Z)
\end{aligned} \tag{9}
$$

Using equations 5 and 6 we get a lower bound on equation 9.

$$
\begin{aligned}
D(p,q) &\geq -(A+W) - \frac{\ln((2\pi)^{2(A+W)}|\Sigma|)}{2} \\
&\quad + E_p(\sum_{ij} (\bar{p}_{ij} - \bar{p}_i \bar{p}_j)(\phi_i - \psi_j)^T (\phi_i - \psi_j)) \\
&\geq -(A+W) - \frac{\ln((2\pi)^{2(A+W)}|\Sigma|)}{2} + E_p(\chi^T \hat{\Lambda} \chi)
\end{aligned}
$$

We get the expression in equation 8 by parameter matching. Differentiating the above equation w.r.t Σ gives us the parameters for the closest Gaussian we project our distribution into.

Derivation of Section 4.4

Now we derive the approximate observation model using Taylor expansion of the exponentiated distance term of the normalization constant, i.e. $e^{-(\phi_i-\psi_j)^T(\phi_i-\psi_j)}$ around parameters ξ_i, ξ_j. We define the gradient (∇) and Hessian (H) for our function. The gradient is defined as follows:

$$\nabla_1(\xi_i, \xi_j) = (\frac{\partial g}{\partial \phi_i})_{\xi_i, \xi_j} = -2e^{-(\xi_i-\xi_j)^T(\xi_i-\xi_j)}(\phi_i - \psi_j)$$

$$\nabla_2(\xi_i, \xi_j) = (\frac{\partial g}{\partial \psi_j})_{\xi_i, \xi_j} = -\nabla_1(\xi_i, \xi_j)$$

$$H = \begin{pmatrix} \frac{\partial^2 g}{\partial \Phi_t^T \partial \Phi_t^T} & \frac{\partial^2 g}{\partial \Psi_t^T \partial \Phi_t} \\ \frac{\partial^2 g}{\partial \Phi_t^T \partial \Psi_t} & \frac{\partial^2 g}{\partial \Psi_t^T \partial \Psi_t^T} \end{pmatrix}_{\xi_i, \xi_j}$$

$$= \begin{pmatrix} H_{11} & H_{12} \\ H_{21} & H_{22} \end{pmatrix}$$

The second order approximation of $e^{-(\phi_i-\psi_j)^T(\phi_i-\psi_j)}$ gives

$$1 + \phi_i^T \nabla_1 + \psi_j^T \nabla_2 + \frac{1}{2}[\Phi_t^T \Psi_t^T] H(\xi_i, \xi_j)[\Phi_t \Psi_t]$$

$$= 1 + \frac{1}{2}[\phi_i^T H_{11}\phi_i + \psi_j^T H_{21}\phi_i + \phi_i^T H_{12}\psi_j + \psi_j^T H_{22}\psi_j] \tag{10}$$

Where $H(\xi_i, \xi_j)$ is H evaluated at ξ_i, ξ_j. For our purpose these values evaluate to the following:

$$\begin{aligned} H_{11} &= 2e^{-(\xi_i-\xi_j)^T(\xi_i-\xi_j)}(2(\xi_i - \xi_j)(\xi_i - \xi_j)^T - I) \\ H_{12} &= -H_{11} \\ H_{21} &= -H_{11}^T \\ H_{22} &= H_{22} \end{aligned} \tag{11}$$

We also define the following symmetric matrix $\overline{\eta}$ and $\overline{\Lambda}$ for making the derivations simple. Also here $\overline{\eta}$ is $2(A+W)$ a dimensional vector and $\overline{\Lambda}$ is a $2(A+W), 2(A+W)$ dimensional symmetric matrix. By i we denote author i and by j we index word j.

$$\begin{aligned} \overline{\eta}_i &= \overline{p}_i \sum_j \overline{p}_j \nabla_1(\xi_i, \xi_j) \\ \overline{\eta}_j &= \overline{p}_j \sum_i \overline{p}_i \nabla_2(\xi_i, \xi_j) \end{aligned} \tag{12}$$

$$\begin{aligned} \overline{\Lambda}_{ii} &= \overline{p}_i \sum_j \overline{p}_j H_{11}(\xi_i, \xi_j) \\ \overline{\Lambda}_{jj} &= \overline{p}_j \sum_i \overline{p}_i H_{22}(\xi_i, \xi_j) \\ \overline{\Lambda}_{ij} &= \overline{p}_i \overline{p}_j H_{12}(\xi_i, \xi_j) \end{aligned} \tag{13}$$

Now using equations (10), (13) and (11) the expectation of the log normalizing constant under the new distribution becomes:

$$E_p(\sum_{ij} \bar{p}_i \bar{p}_j e^{-(\phi_i - \psi_j)^T (\phi_i - \psi_j)})$$
$$= c + E_p[\sum_i \phi_i^T \eta_i + \sum_j \psi_j^T \eta_j] +$$
$$\tfrac{1}{2} E_p[\sum_i \phi_i^T \bar{\Lambda}_{ii} \phi_i + 2 \sum_{ij} \phi_i^T \bar{\Lambda}_{ij} \psi_i + \sum_j \phi_j^T \bar{\Lambda}_{jj} \phi_j]$$
$$= c + E_p[\chi^T \bar{\eta}] + \tfrac{1}{2} E_p[\chi^T \bar{\Lambda} \chi]$$
$$= c + \mu^T \bar{\eta} + \tfrac{1}{2} Tr((\mu\mu^T + \Sigma)\bar{\Lambda})$$

All terms independent of μ, Σ are combined in the constant term c. Hence the approximation of $D(p, q)$ comes out to be,

$$D(p, q) \approx C - \tfrac{1}{2} ln|\Sigma| + tr((\mu\mu^T + \Sigma))\tilde{\Lambda}) + \lambda \mu^T \bar{\eta} +$$
$$\tfrac{\lambda}{2} Tr((\mu\mu^T + \Sigma)\bar{\Lambda})$$

A derivative w.r.t Σ and μ yields

$$\Lambda = \Sigma^{-1} = 2(\tilde{\Lambda} + \tfrac{\lambda}{2}\bar{\Lambda})$$
$$\eta = -\lambda\bar{\eta}$$

which are the required parameters for the Gaussian approximation of the observation model used in the Kalman filter.

Discovering Functional Communities in Dynamical Networks

Cosma Rohilla Shalizi[1], Marcelo F. Camperi[2], and Kristina Lisa Klinkner[1]

[1] Statistics Department, Carnegie Mellon University, Pittsburgh, PA 15213 USA
{cshalizi, klinkner}@cmu.edu
[2] Physics Department, University of San Francisco, San Francisco, CA 94118 USA
camperi@usfca.edu

Abstract. Many networks are important because they are substrates for dynamical systems, and their pattern of functional connectivity can itself be dynamic — they can functionally reorganize, even if their underlying anatomical structure remains fixed. However, the recent rapid progress in discovering the community structure of networks has overwhelmingly focused on that constant anatomical connectivity. In this paper, we lay out the problem of discovering *functional communities*, and describe an approach to doing so. This method combines recent work on measuring information sharing across stochastic networks with an existing and successful community-discovery algorithm for weighted networks. We illustrate it with an application to a large biophysical model of the transition from beta to gamma rhythms in the hippocampus.

1 Introduction

The *community discovery* problem for networks is that of splitting a graph, representing a group of interacting processes or entities, into sub-graphs (communities) which are somehow modular, so that the nodes belonging to a given sub-graph interact with the other members more strongly than they do with the rest of the network. As the word "community" indicates, the problem has its roots in the study of social structure and cohesion [1,2,3], but is related to both general issues of clustering in statistical data mining [4] and to the systems-analysis problem of decomposing large systems into weakly-coupled sub-systems [5,6,7].

The work of Newman and Girvan [8] has inspired a great deal of research on statistical-mechanical approaches to community detection in complex networks. (For a recent partial review, see [9].) To date, however, this tradition has implicitly assumed that the network is defined by persistent, if not static, connections between nodes, whether through concrete physical channels (e.g., electrical power grids, nerve fibers in the brain), or through enduring, settled patterns of interaction (e.g., friendship and collaboration networks). However, networks can also be defined through *coordinated behavior*, and the associated sharing of dynamical information; neuroscience distinguishes these as, respectively, "anatomical"

E.M. Airoldi et al. (Eds.): ICML 2006 Ws, LNCS 4503, pp. 140–157, 2007.
© Springer-Verlag Berlin Heidelberg 2007

and "functional" connectivity [10,11]. The two sorts of connectivity do not map neatly onto each other, and it would be odd if functional modules always lined up with anatomical ones. Indeed, the same system could have many different sets of functional communities in different dynamical regimes. For an extreme case, consider globally-coupled map lattices [12], where the "anatomical" network is fully connected, so there is only a single (trivial) community. Nonetheless, in some dynamical regimes they spontaneously develop many functional communities, i.e., groups of nodes which are internally coherent but with low inter-group coordination [14].

Coupled map lattices are mathematical models, but the distinction between anatomical and functional communities is not merely a conceptual possibility. Observation of neuronal networks *in vivo* show that it is fairly common for, e.g., central pattern generators to change their functional organization considerably, depending on which pattern they are generating, while maintaining a constant anatomy [15]. Similarly, neuropsychological evidence has long suggested that there is no one-to-one mapping between higher cognitive functions and specialized cortical modules, but rather that the latter participate in multiple functions and vice versa, re-organizing depending on the task situation [16]. Details of this picture, of specialized anatomical regions supporting multiple patterns of functional connectivity, have more recently been filled in by brain imaging studies [11]. Similar principles are thought to govern the immune response, cellular signaling, and other forms of biological information processing [17]. Thus, in analyzing these biological networks, it would be highly desirable to have a way of detecting functional communities, rather than just anatomical ones. Similarly, while much of the work on social network organization concerns itself with the persistent ties which are analogous to anatomy, it seems very likely [18,19] that these communities cut in complicated ways across the functional ones defined by behavioral coordination [20,21] or information flow [22]. This is perhaps particularly true of modern societies, which are thought, on several grounds [23,24,19] to be more flexibly organized than traditional ones.

In this paper, we propose a two-part method to discover functional communities in network dynamical systems. Section 2.1 describes the first part, which is to calculate, across the whole of the network, an appropriate measure of behavioral coordination or information sharing; we argue that informational coherence, introduced in our prior work [25], provides such a measure. Section 2.2 describes the other half of our method, using our measure of coordination in place of a traditional adjacency matrix in a suitable community-discovery algorithm. Here we employ the Potts model procedure proposed by Reichardt and Bornholdt [26,27]. Section 2.3 summarizes the method and clarifies the meaning of the functional communities it finds. Section 3 applies our method to a detailed biophysical model of collective oscillations in the hippocampus [28], where it allows us to detect the functional re-organization accompanying the transition from gamma to beta rhythms. Finally, Sect. 5 discusses the limitations of our method and its relations to other approaches (Sect. 5.1) and some issues for future work (Sect. 5.2).

2   Discovering Behavioral Communities

There are two parts to our method for finding functional communities. We first calculate a measure of the behavioral coordination between all pairs of nodes in the network: here, the informational coherence introduced in [25]. We then feed the resulting matrix into a suitable community-discovery algorithm, in place of the usual representation of a network by its adjacency matrix. Here, we have used the Reichardt-Bornholdt algorithm [26], owing to its Hamiltonian form, its ability to handle weighted networks, and its close connection to modularity.

2.1   Informational Coherence

We introduced *informational coherence* in [25] to measure the degree to which the behavior of two systems is coordinated, i.e., how much dynamically-relevant information they share. Because of its centrality to our method, we briefly recapitulate the argument of that paper.

The starting point is the strong notion of "state" employed in physics and dynamical systems theory: the state of the system is a variable which determines the distribution of all present and future observables. In inferential terms, the state is a minimal sufficient statistic for predicting future observations [29], and can be formally constructed as measure-valued process giving the distribution of future events conditional on the history of the process. As a consequence, the state always evolves according to a homogeneous Markov process [30,29].

In a dynamical network, each node i has an associated time-series of observations $X_i(t)$. This is in turn generated by a Markovian state process, $S_i(t)$, which forms its optimal nonlinear predictor. For any two nodes i and j, the informational coherence is

$$IC_{ij} \equiv \frac{I[S_i; S_j]}{\min H[S_i], H[S_j]} \tag{1}$$

where $I[S_i; S_j]$ is the mutual information shared by S_i and S_j, and $H[S_i]$ is the self-information (Shannon entropy) of S_i. Since $I[S_i; S_j] \leq \min H[S_i], H[S_j]$, this is a symmetric quantity, normalized to lie between 0 and 1 inclusive. The construction of the predictive states ensures that $S_i(t)$ encapsulates all information in the past of $X_i(t)$ which is relevant to its future, so a positive value for $I[S_i; S_j]$ means that $S_j(t)$ contains information about the future of $X_i(t)$. That is, a positive value of $I[S_i; S_j]$ is equivalent to the sharing of *dynamically relevant* information between the nodes, manifesting itself as coordinated behavior on the part of nodes i and j.

Clearly, a crucial step in calculating informational coherence is going from the observational time series $X_i(t)$ to the predictive state series $S_i(t)$. In certain cases with completely specified probability models, this can be done analytically [29,31]. In general, however, we are forced to reconstruct the appropriate state-space structure from the time series itself. State reconstruction for deterministic systems is based on the Takens embedding theorem, and is now routine [32]. However, biological and social systems are hardly ever deterministic at experimentally-accessible levels of resolution, so we need a stochastic state reconstruction algorithm. Several exist; we use the CSSR algorithm introduced in [33], since, so far as we know, it is currently the only stochastic state reconstruction algorithm which has been proved statistically consistent (for conditionally stationary discrete sequences). We briefly describe CSSR in Appendix A.

Informational coherence is not, of course, the only possible way of measuring behavioral coordination, or functional connectivity. However, it has a number of advantages over rival measures [25]. Unlike measures of strict synchronization, which insist on units doing exactly the same thing at exactly the same time, it accommodates phase lags, phase locking, chaotic synchronization, etc., in a straightforward and uniform manner. Unlike cross-covariance, or the related spectral coherence, it easily handles nonlinear dependencies, and does not require the choice of a particular lag (or frequency, for spectral coherence), because the predictive states summarize the entire relevant portion of the history. Generalized synchrony measures [34] can handle nonlinear relationships among states, but inappropriately assume determinism. Finally, mutual information among the observables, $I[X_i; X_j]$, can handle nonlinear, stochastic dependencies, but suffers, especially in neural systems, because what we really want to detect are coordinated *patterns* of behavior, rather than coordinated instantaneous actions. Because each predictive state corresponds to a unique statistical pattern of behavior, mutual information among these states is the most natural way to capture functional connectivity.

2.2   The Reichardt-Bornholdt Community Discovery Algorithm

The Reichardt-Bornholdt [26,27] community discovery algorithm finds groups of nodes that are densely coupled to one another, but only weakly coupled to the rest of the network, by establishing a (fictitious) spin system on the network, with a Hamiltonian with precisely the desired properties, and then minimizing the Hamiltonian through simulated annealing. More concretely, every node i is assigned a "spin" σ_i, which is a discrete variable taking an integer value from 1 to a user-defined q. A "community" or "module" will consist of all the nodes with a common spin value. The spin Hamiltonian combines a ferromagnetic term, which favors linked nodes taking the same spin (i.e., being in the same community), and an anti-ferromagnetic term, which favors non-linked nodes taking different spins (i.e., being in different community). Both interactions are of the Potts model type, i.e., they are invariant under permutations of the integers labeling the clusters. After some algebraic manipulation [27], one arrives at the Hamiltonian

$$\mathcal{H}(\sigma) = -\sum_{i \neq j} (A_{ij} - \gamma p_{ij}) \delta(\sigma_i, \sigma_j) \tag{2}$$

where A_{ij} is the adjacency matrix, $\delta(\cdot, \cdot)$ is the Kronecker delta function, p_{ij} is a matrix of non-negative constants giving the relative weights of different possible links, and γ gives the relative contribution of link absence to link presence. The choice of p_{ij} is actually fairly unconstrained, but previous experience with community discovery suggests that very good results are obtained by optimizing the Newman modularity Q [35]

$$Q(\sigma) = \frac{1}{2M} \sum_{i,j} \left(A_{ij} - \frac{k_i k_j}{2M} \right) \delta(\sigma_i, \sigma_j) \tag{3}$$

where k_i is the degree of node i, and $M = \sum_i k_i$ the total number of links. Essentially, Newman's Q counts the number of edges within communities, minus the number which would be expected in a randomized graph where each node preserved its actual degree [9], and σ_i were IID uniform. Setting $p_{ij} = k_i k_j / 2M$ and $\gamma = 1$, we see that $\mathcal{H}(\sigma)$ and $-Q(\sigma)$ differ only by a term (the diagonal part of the sum for Q) which does not depend on the assignment of nodes to communities. Thus, minimizing $\mathcal{H}(\sigma)$ is the same as maximizing the modularity. Varying γ, in this scheme, effectively controls the trade-off between having many small communities and a few large ones [27], and makes it possible to discover a hierarchical community structure, which will be the subject of future work.

While this procedure was originally developed for the case where A_{ij} is a 0-1 adjacency matrix, it also works perfectly well when links take on (positive) real-valued strengths. In particular, using $A_{ij} = IC_{ij}$, we can still maximize the modularity, taking the "degree" of node i to be $k_i = \sum_j IC_{ij}$ [27]. The interpretation of the modularity is now the difference between the strength of intra-community links, and a randomized model where each node shares its link strength indifferently with members of its own and other communities.

2.3 Summary of the Method

Let us briefly summarize the method for discovering functional communities. We begin with a network, consisting of N nodes. For each node, we have a discrete-value, discrete-time ("symbolic") time series, $\{x_i(t)\}$, recorded simultaneously over all nodes. The CSSR algorithm is applied to each node's series separately, producing a set of predictive states for that node, and a time series of those states, $\{s_i(t)\}$. We then calculate the complete set of pairwise informational coherence values, $\{IC_{ij}\}$, using Eq. 1. This matrix is fed into the Reichardt-Bornholdt procedure, with $A_{ij} = IC_{ij}$, which finds an assignment of spins to nodes, $\{\sigma_i\}$, minimizing the Hamiltonian 2. The functional communities of the dynamical network consist of groups of nodes with common spin values. Within each community, the average pairwise coherence of the nodes is strictly greater than would be expected from a randomizing null model (as described in the previous paragraph). Furthermore, between any two communities, the average pairwise coherence of their nodes is strictly less than expected from randomization [27].

3 Test on a Model System of Known Structure: Collective Oscillations in the Hippocampus

We use simulated data as a test case, to validate the general idea of our method, because it allows us to work with a substantial network where we nonetheless have a strong idea of what appropriate results should be. Because of our ultimate concern with the functional re-organization of the brain, we employed a large, biophysically-detailed neuronal network model, with over 1000 simulated neurons.

The model, taken from [28], was originally designed to study episodes of gamma (30–80Hz) and beta (12–30Hz) oscillations in the mammalian nervous system, which often occur successively with a spontaneous transition between them. More concretely, the rhythms studied were those displayed by *in vitro* hippocampal (CA1) slice preparations and by *in vivo* neocortical EEGs.

The model contains two neuron populations: excitatory (AMPA) pyramidal neurons and inhibitory ($GABA_A$) interneurons, defined by conductance-based Hodgkin-Huxley-style equations. Simulations were carried out in a network of 1000 pyramidal cells and 300 interneurons. Each cell was modeled as a one-compartment neuron with all-to-all coupling, endowed with the basic sodium and potassium spiking currents, an external applied current, and some Gaussian input noise. The anatomical, synaptic connections were organized into blocks, as shown in Fig. 2.

The first 10 seconds of the simulation correspond to the gamma rhythm, in which only a group of neurons is made to spike via a linearly increasing applied current. The beta rhythm (subsequent 10 seconds) is obtained by activating pyramidal-pyramidal recurrent connections (potentiated by Hebbian preprocessing as a result of synchrony during the gamma rhythm) and a slow outward after-hyper-polarization (AHP) current (the M-current), suppressed during gamma due to the metabotropic activation used in the generation of the rhythm. During the beta rhythm, pyramidal cells, silent during gamma rhythm, fire on a subset of interneurons cycles (Fig. 1).

4 Results on the Model

A simple heat-map display of the informational coherence (Fig. 3) shows little structure among the active neurons in either regime. However, visual inspection of the rastergrams (Fig. 1) leads us to suspect the presence of two very large functional communities: one, centered on the inhibitory interneurons and the excitatory pyramidal neurons most tightly coupled to them, and another of the more peripheral excitatory neurons. During the switch from the gamma to the beta rhythm, we expect these groups to re-organize.

These expectations are abundantly fulfilled (Fig. 4). We identified communities by running the Reichardt-Bornholdt algorithm with the maximum number of communities (spin states) set to 25, the modularity Hamiltonian, and $\gamma = 1$. (Results were basically unchanged at 40 or 100 spin values.) In both regimes, there are two overwhelmingly large communities, containing almost all of the neurons which actually fired, and a handful of single-neuron communities. The significant change, visible in the figure, is in the organization of these communities.

During the gamma rhythm, the 300 interneurons form the core of the larger of these two communities, which also contains 199 pyramidal neurons. Another 430

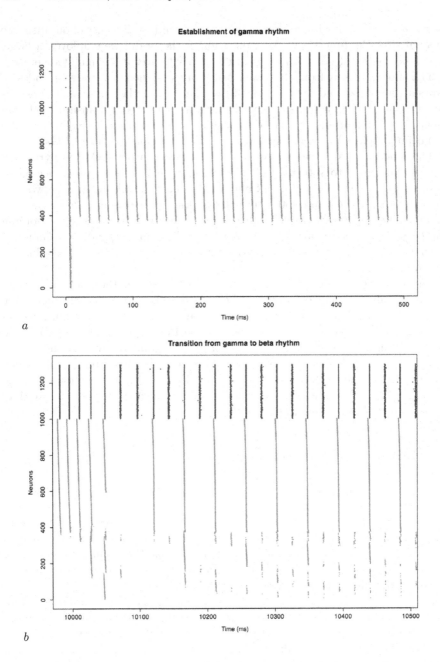

a

b

Fig. 1. Rastergrams of neuronal spike-times in the network. Excitatory, pyramidal neurons (numbers 1 to 1000) are green, inhibitory interneurons (numbers 1001 to 1300) are red. During the first 10 seconds (*a*), the current connections among the pyramidal cells are suppressed and a gamma rhythm emerges (left). At $t = 10$s, those connections become active, leading to a beta rhythm (*b*, right).

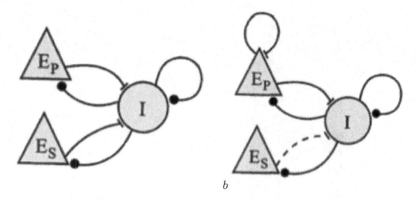

Fig. 2. Schematic depiction of the anatomical network. Here nodes represent populations of cells: excitatory pyramidal neurons (triangles labeled E) or inhibitory interneurons (large circle labeled I). Excitatory connections terminate in bars, inhibitory connections in filled circles. During the gamma rhythm (*a*), the pyramidal neurons are coupled to each other only indirectly, via the interneurons, and dynamical effects separate the pyramidal population into participating (E_P) and suppressed (E_S) subpopulations. During the beta rhythm (*b*), direct connections among the E_P neurons, built up, but not activated, by Hebbian learning under the gamma rhythm are turned on, and the connection from the E_S neurons to the interneurons are weakened by the same Hebbian process (dashed line).

pyramidal neurons belong to a second community. A final 5 pyramidal cells are in single-neuron communities; the rest do not fire at all. A hierarchical analysis (not shown) has the two large communities merging into a single super-community. The regular alternation of the two communities among the pyramidal neurons, evident in Fig. 4*a*, is due to the fact that the external current driving the pyramidal neurons is not spatially uniform.

With the switch to the beta rhythm, the communities grow and re-organize. The community centered on the interneurons expands, to 733 neurons, largely by incorporating many low-index pyramidal neurons which had formerly been silent, and are now somewhat erratically synchronized, into its periphery. Interestingly, many of the latter are only weakly coherent with any one interneuron (as can be seen by comparing Figs. 3*b* and 4*b*). What is decisive is rather their stronger over-all pattern of coordination with the interneurons, shown by sharing a common (approximate) firing period, which is half that of the high-index pyramidal cells (Fig. 1*b*). Similarly, the other large community, consisting exclusively of pyramidal neurons, also grows (to 518 members), again by expanding into the low-index part of the network; there is also considerable exchange of high-index pyramidal cells between the two communities. Finally, nine low-index neurons, which fire only sporadically, belong in clusters of one or two cells.

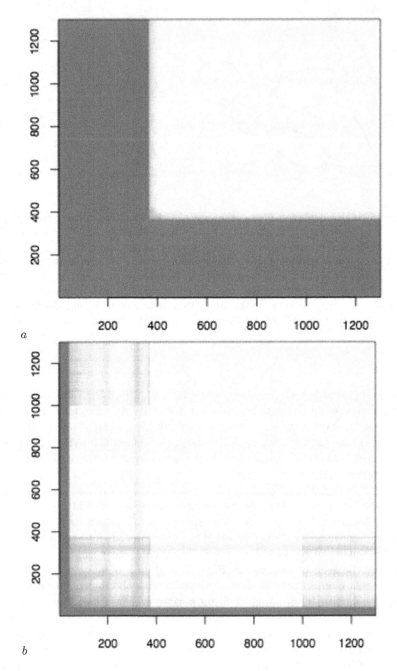

a

b

Fig. 3. Heat-maps of coordination across neurons in the network, measured by informational coherence. Colors run from red (no coordination) through yellow to pale cream (maximum).

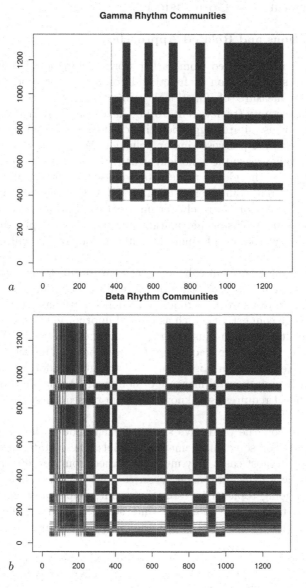

Fig. 4. Division of the network into functional communities. Black points denote pairs of nodes which are both members of a given community. During the gamma rhythm (a) the interneurons (numbered 1001 to 1300) form the core of a single community, along with some of the active pyramidal neurons; because of the spatially modulated input received by the latter, however, some of them belong to another community. During beta rhythm (b), the communities re-organize, and in particular formerly inactive pyramidal neurons are recruited into the community centered on the interneurons, as suggested by the rastergrams.

5   Discussion and Conclusion

5.1   Limitations and Related Approaches

Our method is distinguished from earlier work on functional connectivity primarily by our strong notion of functional community or module, and secondarily by our measure of functional connectivity. Previous approaches to functional connectivity (reviewed in [10,11]) have either have no concept of functional cluster, or use simple agglomerative clustering [4]; their clusters are just groups of nodes with pairwise-similar behavior. We avoid agglomerative clustering for the same reason it is no longer used to find anatomical communities: it is insensitive to the global pattern of connectivity, and fails to divide the network into coherent components. Recall (Sect. 2.3) that every functional community we find has more intra-cluster information sharing than is expected by chance, and less inter-cluster information sharing. This is a plausible formalization of the intuitive notion of "module", but agglomeration will not, generally, deliver it.

As for using informational coherence to measure functional connectivity, we discussed its advantages over other measures in Sect. 2.1 above, and at more length in [25]. Previous work on functional connectivity has mostly used surface features to gauge connectivity, such as mutual information between observables. (Some of the literature on clustering general time series, e.g. [36,37,38], uses hidden Markov models to extract latent features, but in a mixture-model framework very different from our approach.) The strength of informational coherence is that it is a domain-neutral measure of nonlinear, stochastic coordination; its weakness is that it requires us to know the temporal sequence of predictive states of all nodes in the network.

Needing to know the predictive states of each node is the major limitation of our method. For some mathematical models, these states are analytically calculable, but in most cases they must be learned from discrete-value, discrete-time ("symbolic") time series. Those series must be fairly long; exactly how long is an on-going topic of investigation,[1] but, empirically, good results are rare with less than a few thousand time steps. Similarly, reliable estimates of the mutual information and informational coherence also require long time series.

Predictive states can be mathematically defined for continuous-value, continuous-time systems [30], but all current algorithms for discovering them, not just CSSR, require symbolic time series. (Devising a state-reconstruction procedure for continuous systems is another topic of ongoing research.) Spike trains, like e-mail networks [22], are naturally discrete, so this is not an issue for them, but in most other cases we need to find a good symbolic partition first, which is non-trivial [39]. The need for long symbolic time series may be especially difficult to meet with social networks.

---

[1] CSSR converges on the true predictive states (see the appendix), but the rate of convergence is not yet known.

5.2 Directions for Future Work

Our results on the model of hippocampal rhythms, described in the previous section, are quite promising: our algorithm discovers functional communities whose organization and properties make sense, given the underlying micro-dynamics of the model. This suggests that it is worthwhile to apply the method to systems where we lack good background knowledge of the functional modules. Without pre-judging the results of those investigations, however, we would like to highlight some issues for future work.

1. Our method needs the full matrix of informational coherences, which is an $O(N^2)$ computation for a network of size N. If we are interested in the organization of only part of the network, can we avoid this by defining a local community structure, as was done for anatomical connectivity by [40]? Alternatively, if we know the anatomical connectivity, can we restrict ourselves to calculating the informational coherence between nodes which are anatomically tied? Doing so with our model system led to basically the same results (not shown), which is promising; but in many real-world systems the anatomical network is itself uncertain.

2. The modularity Hamiltonian of Sect. 2.2 measures how much information each node shares with other members of its community on a pairwise basis. However, some of this information could be redundant across pairs. It might be better, then, to replace the sum over pairs with a higher-order coherence. The necessary higher-order mutual informations are easily defined [10,41,42], but the number of measurements needed to estimate them from data grows exponentially with the number of nodes. However, it may be possible to approximate them using the same Chow-Liu bounds employed by [25] to estimate the global coherence.

3. It would be good if our algorithm did not simply report a community structure, but also assessed the likelihood of the same degree of modularity arising through chance, i.e., a significance level. For anatomical communities, Guimera et al. [43] exploit the spin-system analogy to show that random graph processes without community structure will nonetheless often produce networks with non-zero modularity, and (in effect) calculate the sampling distribution of Newman's Q using both Erdös-Rényi and scale-free networks as null models. (See however [44] for corrections to their calculations.) To do something like this with our algorithm, we would need a null model of functional communities. The natural null model of functional connectivity is simply for the dynamics at all nodes to be independent, and (because the states are Markovian) it is easy to simulate from this null model and then bootstrap p-values. We do not yet, however, have a class of dynamical models where there nodes share information, but do so in a completely distributed, a-modular way.

4. A variant of the predictive-state analysis that underlies informational coherence is able to identify coherent structures produced by spatiotemporal dynamics [45]. Moreover, these techniques can be adapted to network dynamics, if the anatomical connections are known. This raises numerous questions. Are functional communities also coherent structures? Are coherent structures in

networks [46] necessarily functional communities? Can the higher-order interactions of coherent structures in regular spatial systems be ported to networks, and, if so, could functional re-organization be described as a dynamical process at this level?

5.3   Conclusion

Network dynamical systems have both anatomical connections, due to persistent physical couplings, and functional ones, due to coordinated behavior. These are related, but logically distinct. There are now many methods for using a network's anatomical connectivity to decompose it into highly modular communities, and some understanding of these methods' statistical and statistical-mechanical properties. The parallel problem, of using the pattern of functional connectivity to find functional communities, has scarcely been explored. It is in many ways a harder problem, because measuring functional connectivity is harder, and because the community organization is itself variable, and this variation is often more interesting than the value at any one time.

In this paper, we have introduced a method of discovering functional modules in stochastic dynamical networks. We use informational coherence to measure functional connectivity, and combine this with a modification of the Potts-model community-detection procedure. Our method gives good results on a biophysical model of hippocampal rhythms. It divides the network into two functional communities, one of them based on the inhibitory interneurons, the other consisting exclusively of excitatory pyramidal cells. The two communities change in relative size and re-organize during the switch from gamma to beta rhythm, in ways which make sense in light of the underlying model dynamics. While there are theoretical issues to explore, our success on a non-trivial simulated network leads us to hope that we have found a general method for discovering functional communities in dynamic networks.

Acknowledgments

Thanks to L. A. N. Amaral, S. Bornholdt, A. Clauset, R. Haslinger, C. Moore and M. E. J. Newman for discussing community discovery and/or network dynamics; to J. Reichardt for valuable assistance with the implementation of his algorithm and general suggestions; and to the editors for their patience.

References

1. Simmel, G.: Conflict and the Web of Group Affiliation. The Free Press, Glencoe, Illinois (1955) Translated by Kurt H. Wolff and Reinhard Bendix, with a foreword by Everett C. Hughes.
2. Scott, J.: Social Network Analysis: A Handbook. Sage Publications, Thousand Oaks, California (2000)
3. Moody, J., White, D.R.: Social cohesion and embeddedness: A hierarchical conception of social groups. American Sociological Review **68** (2003) 103–127

4. Hand, D., Mannila, H., Smyth, P.: Principles of Data Mining. MIT Press, Cambridge, Massachusetts (2001)
5. Simon, H.A.: The architecture of complexity: Hierarchic systems. Proceedings of the American Philosophical Society **106** (1962) 467–482 Reprinted as chapter 8 of [53].
6. Alexander, C.: Notes on the Synthesis of Form. Harvard University Press, Cambridge, Massachusetts (1964)
7. Šijlak, D.S.: Decentralized Control of Complex Systems. Academic Press, Boston (1990)
8. Newman, M.E.J., Girvan, M.: Finding and evaluating community structure in networks. Physical Review E **69** (2003) 026113
9. Newman, M.E.J.: Finding community structure in networks using the eigenvectors of matrices. E-print, arxiv.org, physics/0605087 (2006)
10. Sporns, O., Tononi, G., Edelman, G.M.: Theoretical neuroanatomy: Relating anatomical and functional connectivity in graphs and cortical connection matrices. Cerebral Cortex **10** (2000) 127–141
11. Friston, K.: Beyond phrenology: What can neuroimaging tell us about distributed circuitry? Annual Review of Neuroscience **25** (2002) 221–250
12. Kaneko, K.: Clustering, coding, switching, hierarchical ordering, and control in a network of chaotic elements. Physica D **41** (1990) 137–172
13. Chazottes, J.R., Fernandez, B., eds.: Dynamics of Coupled Map Lattices and of Related Spatially Extended Systems. Volume 671 of Lecture Notes in Physics. Springer-Verlag, Berlin (2005)
14. Delgado, J., Solé, R.V.: Characterizing turbulence in globally coupled maps with stochastic finite automata. Physics Letters A **314** (2000) 314–319
15. Selvertson, A.I., Moulins, M., eds.: The Crustacean Stomatogastric System: A Model for the Study of Central Nervous Systems. Springer-Verlag, Berlin (1987)
16. Luria, A.R.: The Working Brain: An Introduction to Neuropsychology. Basic Books, New York (1973)
17. Segel, L.A., Cohen, I.R., eds.: Design Principles for the Immune System and Other Distributed Autonomous Systems. Oxford University Press, Oxford (2001)
18. Chisholm, D.: Coordination without Hierarchy: Informal Structures in Multiorganizational Systems. University of California Press, Berkeley (1989)
19. Luhmann, N.: Social Systems. Stanford University Press, Stanford (1984/1995) Translated by John Bednarz, Jr., with Dirk Baecker. Foreword by Eva M. Knodt. First published as *Soziale Systeme: Grundriss einer allgemeinen Theorie*, Frankfurt am Main: Suhrkamp-Verlag.
20. Lindblom, C.E.: The Intelligence of Democracy: Decision Making through Mutual Adjustment. Free Press, New York (1965)
21. Young, H.P.: Individual Strategy and Social Structure: An Evolutionary Theory of Institutions. Princeton University Press, Princeton (1998)
22. Eckmann, J.P., Moses, E., Sergi, D.: Entropy of dialogues creates coherent structures in e-mail traffic. Proceedings of the National Academy of Sciences (USA) **101** (2004) 14333–14337
23. Coser, R.L.: In Defense of Modernity: Role Complexity and Individual Autonomy. Stanford University Press, Stanford, California (1991)
24. Gellner, E.: Plough, Sword and Book: The Structure of Human History. University of Chicago Press, Chicago (1988)
25. Klinkner, K.L., Shalizi, C.R., Camperi, M.F.: Measuring shared information and coordinated activity in neuronal networks. In Weiss, Y., Schölkopf, B., Platt, J.C., eds.: Advances in Neural Information Processing Systems 18 (NIPS 2005), Cambridge, Massachusetts, MIT Press (2006) 667–674

26. Reichardt, J., Bornholdt, S.: Detecting fuzzy community structures in complex networks with a Potts model. Physical Review Letters **93** (2004) 218701
27. Reichardt, J., Bornholdt, S.: Statistical mechanics of community detection. Physical Review E **74** (2006) 016110
28. Olufsen, M.S., Whittington, M.A., Camperi, M., Kopell, N.: New roles for the gamma rhythm: Population tuning and preprocessing for the beta rhythm. Journal of Computational Neuroscience **14** (2003) 33–54
29. Shalizi, C.R., Crutchfield, J.P.: Computational mechanics: Pattern and prediction, structure and simplicity. Journal of Statistical Physics **104** (2001) 817–879
30. Knight, F.B.: A predictive view of continuous time processes. Annals of Probability **3** (1975) 573–596
31. Littman, M.L., Sutton, R.S., Singh, S.: Predictive representations of state. In Dietterich, T.G., Becker, S., Ghahramani, Z., eds.: Advances in Neural Information Processing Systems 14, Cambridge, Massachusetts, MIT Press (2002) 1555–1561
32. Kantz, H., Schreiber, T.: Nonlinear Time Series Analysis. Cambridge University Press, Cambridge, England (1997)
33. Shalizi, C.R., Klinkner, K.L.: Blind construction of optimal nonlinear recursive predictors for discrete sequences. In Chickering, M., Halpern, J., eds.: Uncertainty in Artificial Intelligence: Proceedings of the Twentieth Conference, Arlington, Virginia, AUAI Press (2004) 504–511
34. Quian Quiroga, R., Kraskov, A., Kreuz, T., Grassberger, P.: Performance of synchronization measures in real data: case studies on electroencephalographic signals. Physical Review E **65** (2002) 041903
35. Newman, M.E.J.: Mixing patterns in networks. Physical Review E **67** (2003) 026126
36. Smyth, P.: Clustering sequences using hidden Markov models. In Mozer, M.C., Jordan, M.I., Petsche, T., eds.: Advances in Neural Information Processing 9, Cambridge, Massachusetts, MIT Press (1997) 648–654
37. Oates, T., Firiou, L., Cohen, P.R.: Clustering time series with hidden Markov models and dynamic time warping. In Giles, C.L., Sun, R., eds.: Proceedings of the IJCAI-99 Workshop on Neural, Symbolic and Reinforcement Learning Methods for Sequence Learning. (1999) 17–21
38. Cadez, I.V., Gaffney, S., Smyth, P.: A General Probabilistic Framework for Clustering Individuals and Objects. In Ramakrishnan, R., Stolfo, S., Bayardo, R., Parsa, I., eds.: Proceedings of the sixth ACM SIGKDD international conference on Knowledge discovery and data mining, New York, ACM Press (2000) 140–149
39. Hirata, Y., Judd, K., Kilminster, D.: Estimating a generating partition from observed time series: Symbolic shadowing. Physical Review E **70** (2004) 016215
40. Clauset, A.: Finding local community structure in networks. Physical Review E **72** (2005) 026132
41. Amari, S.i.: Information geometry on hierarchy of probability distributions. IEEE Transactions on Information Theory **47** (2001) 1701–1711
42. Schneidman, E., Still, S., Berry, M.J., Bialek, W.: Network information and connected correlations. Physical Review Letters **91** (2003) 238701
43. Guimera, R., Sales-Pardo, M., Amaral, L.A.N.: Modularity from fluctuations in random graphs and complex networks. Physical Review E **70** (2004) 025101
44. Reichardt, J., Bornholdt, S.: When are networks truly modular? E-print, arxiv.org, cond-mat/0606220 (2006)
45. Shalizi, C.R., Haslinger, R., Rouquier, J.B., Klinkner, K.L., Moore, C.: Automatic filters for the detection of coherent structure in spatiotemporal systems. Physical Review E **73** (2006) 036104

46. Moreira, A.A., Mathur, A., Diermeier, D., Amaral, L.A.N.: Efficient system-wide coordination in noisy environments. Proceedings of the National Academy of Sciences (USA) **101** (2004) 12085–12090
47. Varn, D.P., Crutchfield, J.P.: From finite to infinite range order via annealing: The causal architecture of deformation faulting in annealed close-packed crystals. Physics Letters A **324** (2004) 299–307
48. Ray, A.: Symbolic dynamic analysis of complex systems for anomaly detection. Signal Processing **84** (2004) 1115–1130
49. Padró, M., Padró, L.: Applying a finite automata acquisition algorithm to named entity recognition. In: Proceedings of 5th International Workshop on Finite-State Methods and Natural Language Processing (FSMNLP'05). (2005)
50. Padró, M., Padró, L.: A named entity recognition system based on a finite automata acquisition algorithm. Procesamiento del Lenguaje Natural **35** (2005) 319–326
51. Padró, M., Padró, L.: Approaching sequential NLP tasks with an automata acquisition algorithm. In: Proceedings of International Conference on Recent Advances in NLP (RANLP'05). (2005)
52. Kullback, S.: Information Theory and Statistics. 2nd edn. Dover Books, New York (1968)
53. Simon, H.A.: The Sciences of the Artificial. Third edn. MIT Press, Cambridge, Massachusetts (1996) First edition 1969.

A   The CSSR Algorithm

This appendix briefly describes the CSSR algorithm we use to reconstruct the effective internal states of each node in the network. For details, see [33]; for an open-source C++ implementation, see http://bactra.org/CSSR/. For recent applications of the algorithm to problems in crystallography, anomaly detection and natural language processing, see [47,48,49,50,51].

We wish to predict a dynamical system or stochastic process $\{X_t\}$. By X_s^t we will denote the whole trajectory of the process from time s to time t, inclusive, by X_t^+ the whole "past" or "history" of the process through time t, and by X_t^+ its "future", its trajectory at times strictly greater than t. The "state" of $\{X_t\}$ at time t is a variable, S_t, which fixes the distribution of all present or future observations, i.e., the distribution of $X^+(t)$ [29,31]. As such, the state is a minimal sufficient statistic for predicting the future of the process. Sufficiency is equivalent to the requirement that $I[X_t^+; X_t^-] = I[X_t^+; S_t]$, where $I[\cdot; \cdot]$ is the mutual information [52]. In general, $S_t = \epsilon(X_t^-)$, for some measurable functional $\epsilon(\cdot)$ of the whole past history of the process up to and including time t. If $\{X_t\}$ is Markovian, then ϵ is a function only of X_t, but in general the state will incorporate some history or memory effects. Each state, i.e., possible value of ϵ, corresponds to a predictive distribution over future events, and equally to an equivalence class of histories, all of which lead to that conditional distribution over future events. State-reconstruction algorithms use sample paths of the process to find approximations $\hat{\epsilon}$ to the true minimal sufficient statistic ϵ, and ideally the approximations converge, at least in probability. The CSSR algorithm [33] does so, for discrete-valued, discrete-time, conditionally-stationary processes.

CSSR is based on the following result about predictive sufficiency [29, pp. 842–843]. Suppose that ϵ is next-step sufficient, i.e., $I[X_{t+1}; X_t^-] = I[X_{t+1}; \epsilon(X_t^-)]$, and that it can be updated recursively: for some measurable function T, $\epsilon(X_{t+1}^-) = T(\epsilon(X_t^-, X_{t+1}))$. Then ϵ is predictively sufficient for the whole future of the process — intuitively, the recursive updating lets us chain together accurate next-step predictions to go as far into the future as we like. CSSR approximates ϵ by treating it as a partition, or set of equivalence classes, over histories, and finding the coarsest partition which meets both of the conditions of this theorem. Computationally, CSSR represents states as sets of suffixes, so a history belongs to a state (equivalence class) if it terminates in one of the suffixes in that state's representation. That is, a history, x_t^-, will belong to the class C, $x_t^- \in C$, if $x_{t-|c|+1}^t = c$, for some suffix c assigned to C, where $|c|$ is the length of the suffix.[2]

In the first stage, CSSR tries to find a partition of histories which is sufficient for next-step prediction. It begins with the trivial partition, in which all histories belong to the same equivalence class, defined by the null suffix (corresponding to an IID process), and then successively tests whether longer and longer suffices give rise to the same conditional distribution for the next observation which differ significantly from the class to which they currently belong. That is, for each class C, suffix c in that class, and possible observable value a, it tests whether $\Pr\left(X_{t+1}|X_t^- \in C\right)$ differs from $\Pr\left(X_{t+1}|X_{t-|c|+1}^t = c, X_{t-|c|} = a\right)$. (We use standard tests for discrepancy between sampled distributions.) If an extended, child suffix (ac) does not match its current classes, the parent suffix (c) is deleted from its class (C), and CSSR checks whether the child matches any existing class; if so it is re-assigned to the closest one, and the partition is modified accordingly. Only if a suffix's conditional distribution $(\Pr\left(X_{t+1}|X_{t-l}^t = ac\right))$ differs significantly from all existing classes does it get its own new cell in the partition.

The result of this stage is a partition of histories (i.e., a statistic) which is close to being next-step sufficient, the sense of "close" depending on the significance test. In the second stage, CSSR iteratively refines this partition until it can be recursively updated. This can always be done, though it is potentially the most time-consuming part of the algorithm.[3] The output of CSSR, then, is a set of states which make good next-step predictions and can be updated recursively, and a statistic $\hat{\epsilon}$ mapping histories to these states.

If the true number of predictive states is finite, and some mild technical assumptions hold [33], a large deviations argument shows that $\Pr\left(\hat{\epsilon} \neq \epsilon\right) \to 0$ as the sample size $n \to \infty$. That is, CSSR will converge on the minimal sufficient statistic for the data-generating process, even though it lacks an explicit minimization step. Furthermore, once the right statistic has been discovered, the expected L_1 (total variation) distance between the actual predictive distribution,

---

[2] The algorithm ensures that there are never overlapping suffixes in distinct states.

[3] In terms of automata theory, recursive updating corresponds to being a "deterministic" automaton, and non-deterministic automata always have deterministic equivalents.

$\Pr\left(X_t^+|\epsilon(X_t^-)\right)$ and that forecast by the reconstructed states, $\Pr\left(X_t^-|\hat{\epsilon}(X_t^-)\right)$, goes to zero with rate $O(n^{-1/2})$, which is the same rate as for IID data. The time complexity of the algorithm is at worst $O(n) + O(k^{2L+1})$, where k is the number of discrete values possible for X_t, and L is the maximum length of suffices considered in the reconstruction. Empirically, average-case time complexity is much better than this.

Empirical Analysis of a Dynamic Social Network Built from PGP Keyrings

Robert H. Warren[1], Dana Wilkinson[1], and Mike Warnecke[2]

[1] David R. Cheriton School of Computer Science
University of Waterloo
Waterloo, Canada
{rhwarren,d3wilkin}@uwaterloo.ca
[2] PSW Applied Research Inc.
Waterloo, Canada
mwarnecke@pswappliedresearch.com

Abstract. Social networks are the focus of a large body of research. A number of popular email encryption tools make use of online directories to store public key information. These can be used to build a social network of people connected by email relationships. Since these directories contain creation and expiration time-stamps, the corresponding network can be built and analysed dynamically. At any given point, a snapshot of the current state of the model can be observed and traditional metrics evaluated and compared with the state of the model at other times.

We show that, with this described data set, simple traditional predictive measures do vary with time. Moreover, singular events pertinent to the participants in the social network (such as conferences) can be correlated with or implied by significant changes in these measures. This provides evidence that the dynamic behaviour of social networks should not be ignored, either when analysing a real model or when attempting to generate a synthetic model.

1 Introduction

One of the elements of public key cryptography systems such as Pretty Good Privacy[TM] and GNU Privacy Guard is the need to guarantee the validity and authenticity of public keys. As a solution, key servers dispense key trust information uploaded by key owners in the form of keys signatures. The trust information is inserted into the system based on each users belief that the key that they are signing is the one belonging to the intended user. These key servers are a significant source of historical information as the public keys contain both identity and trust relationships. The common practice of limiting the lifetime of keys and of signatures based on calendar time ensures that stale information can be identified. This allows one to view sets of key-rings within key servers as social networks.

Using the time-stamped data it is possible to trace the entry and departure of persons within the systems as well as the relationships connecting them. At

E.M. Airoldi et al. (Eds.): ICML 2006 Ws, LNCS 4503, pp. 158–171, 2007.

each time it is possible to compute a number of metrics and statistics on the new relationship or on the social network graph as a whole at that point.

We show that these metrics change over the lifetime of the network and that some of the more distinct changes are highly correlated with events relevant to the actors in the network. There was, for example, a distinct increase in the average previous-shortest distance between two newly-connected actors in a Debian mailing list immediately after a Linux conference.

This implies that static metrics are insufficient for analysing and describing the behaviour in this network, and provides general evidence that care must be taken when using only static metrics in analysing other such networks. Such metrics should be recomputed continuously and the temporal differences accounted for. Additionally, these results could be of benefit in modelling "realistic" synthetic social networks.

Further, we demonstrate that this network could, at any given time step, consist of many disconnected components, on the order of the number of nodes in the network. This indicates that care should be taken when using algorithms or techniques which require the assumption that the network is connected, especially since in a dynamically built network components could easily be merging and splitting over time.

The remainder of the paper is organised as follows. We start with a brief summary of graphs and social networks as well as a review of PGP^{TM}. Then, we describe in detail the two data sets that we focus on, one relatively small (extracted from the Debian developers key server) and one much larger (extracted from the U Alberta key server), as well as report some basic statistics on them. Dynamic networks were built up from those data sets so we next describe and report the various metrics and statistics measured throughout the life of these network graphs.

2   Background

2.1   Graphs and Social Networks

Graph theory is old and well-studied, with a plethora of concepts and algorithms. For an introduction to graphs and graph algorithms see, for example, [1]. See [2] for a more comprehensive reference.

Formally, a graph is a pair $G = \{V, E\}$ where V is a set of nodes and E is a set of edges which in turn are pairs of nodes (i.e. $E = \{e = (h, t) : h \in V \text{ and } t \in V\}$). Note that in our definition of a graph (sometimes referred to as an *undirected graph*) the order of the two points associated with an edge is unimportant[1]. Two edges are said to be *joined* if they share the same node between them (i.e. $e_i = (n_1, n_2), e_j = (n_2, n_3)$). A *path* between two nodes is a collection of consecutively joined edges that connect those two nodes. There are a variety of well known algorithms for determining the *shortest path* from one node

---

[1] Contrasted with a *digraph* or *directed graph*, in which an edge is an ordered pair of nodes. The graphs constructed in this document are all *undirected*.

to another. Finally, a *connected component* is a subset of a graph where every pair of nodes in that subset are connected by some path. If a graph is composed of only one connected component then the graph is said to be *connected*.

The idea of social networks is simple—to model social and sociological data using graphs. For a good introduction to social network analysis see [3] or the more recent [4].

Interest in social networks has been around since at least the 1950's. Modelling collections of social actors as nodes in a graph and their relationships as edges provided a paradigm that has since been utilised in a variety of different areas, from studying the neural pathways of bacteria to analysing power grids [5]. In 1967 Milgram [6] formalised the "small world" property that has been found in many social networks. Given any two nodes in a small world, it is highly probable that those nodes are connected by a relatively short path. More recently, Watts' book [7] on the small worlds phenomenon seems to have sparked even more research in the area.

Initially, interest in social networks and small worlds was primarily focused on using the graph paradigm to model and analyse data. More recently, researchers have started looking at various methods of generating synthetic social networks on which a variety of algorithms can be tested. Typically, social networks that have the small world property are desired.

Interest in this small worlds property has translated into interest in a variety of different methods for evaluating a new relationship. Just before an edge is added to the network, the shortest path between the two nodes associated with an edge can be recorded. Presumably, if this previous shortest path is, on average, very low then the network will have the small worlds property (see for example [8]). This leads to a useful tool in analysing social networks, as this metric is often easy to measure. Indeed there are a host of different measures that are associated with such new relationships, all of which are based in some way or another on the concept of measuring the path or paths that exist between two nodes before an edge relating them is added (see for example [9] for a summary of some of these measures).

One thing to note is that some of these measures, as well as other social network algorithms, may require that the network be connected, either to guarantee a performance bound or, in some cases, to work at all.

2.2 PGP™

Pretty Good Privacy (PGP™) and variants (such as GNU Privacy Guard, GPG) are programs for encrypting and signing e-mail. They can be used to encrypt entire e-mail messages but more often are used to sign an e-mail as a way of guaranteeing that the e-mail is actually a product of the person who signed it.

PGP™ uses the RSA (Rivest Shamir Adleman) public and private key crypto-system. Public key methods work by generating separate encryption (public) and decryption (private) keys in such a way that decryption of a message with the public key is nearly impossible. This allows mass distribution of the public key without concern. Anyone can encrypt messages but only someone with the private key can decrypt them.

PGP^{TM} can also be used to apply a digital signature to a message without encrypting it. This is normally used in public postings to allow others to confirm that the message actually came from a particular person. Once a digital signature is created, it is almost impossible for anyone to modify either the message or the signature without the modification being detected by PGP^{TM}.

In order to verify the signature of an e-mail, the public key is needed. Without key servers, people would have to distribute and find these keys themselves. To facilitate this process, key servers store the (public) PGP^{TM} keys and key certificates. Anyone looking for a public PGP^{TM} key can search for and retrieve it from the key servers (The key servers synchronise with each other—if someone adds a key to a key server it is distributed to all key servers).

Initially, a person must actively "sign" the key of another person (indicating that they trust that that key belongs to that person). However, once a person has signed someone's key, that key now becomes trusted by the first person. In this way it is possible to verify the validity of a particular key. A key is only trusted if it is signed.

These chains of signatures build up like a web, called the *web of trust*. This web-like structure is no accident. It is important to have as many disjoint paths as possible to reduce the chance that someone can fake a confirmation chain with a wrong signature.

Everyone who uses PGP^{TM} (or its variants) has a *key-ring* of (mostly) valid public keys. Additionally, a trust value can be assigned to each public key indicating how much a person believes in the authenticity of the key. The validity of a key can be determined by thresholding this trust value. Almost all of this data can be mined from public key servers.

3 The Data

GPG and PGP^{TM} key networks have a number of elements that make them interesting data sources for our purposes; several key analyses have been done in the past on trust relationships within key-rings with an eye at establishing the authenticity of keys and the reliability of the key signing process (e.g., [10]). We pursue here a different approach in that we are not interested in the keys themselves as much as the relationships which they imply between the individuals within the key-ring universe.

The distinction is important in that different individuals may have multiple keys for multiple roles which have not been linked for historical or operational reasons. Hence, while historically the social distance within a group of individuals was calculated with respect to key signatures with authenticity as an objective, we only wish to establish a reasonable expectation that a relationship does exist. We make an implicit assumption that the process used by people to determine key trust is directly linked to the strength of the relationship between the two people and not on a particular relationship between two specific keys.

The keys contain a free form identifier string that is set by the key owner. By a loose convention, this is usually composed of the email address ("John Doe <johndoe@somewhere.com>") of the key owner along with a brief longhand description ("Work place software distribution key").

The keys were then pre-processed to resolve individuals to their public keys, even if an owner-to-owner signature between them was missing. To do this the email labelling data was merged using an m-to-n merge: keys having multiple email addresses were matched with keys labelled with those same email addresses. This ensured that we could obtain an unique identifier for each person within the database.

Signatures between keys were assumed to indicate some form of friendship between individuals. This assumption can be challenged in that key signatures are granted on an opportunistic basis that may not be completely based on friendship, as much as social access. This may explain with some individuals in key networks have a disproportionate 'friend' network that is not reciprocated. While a metric for the level of trust accorded to each key was available, we chose not to make use of it in this research.

Using the time-stamps we then tracked the evolution of the social network from the addition of the first node to the end of the data collection period. There are four possible changes that can occur in the network:

1. Node (person) addition (key creation)
2. Node (person) removal (key expiry/revocation)
3. Edge (relationship) addition (signature creation)
4. Edge (relationship) removal (signature expiry/revocation)

We thus labelled the identifying and friendship data with time-stamps. This was done to prevent stale social information from flooding our analysis network. Individuals and relationships were temporally removed from the dataset where their underlying keys and signatures where cryptographically revoked or expired. Because in the vast majority of cases no expiration date had been set for the keys, we applied a timeout period of one year after the last sign of activity (key creation or signing) from the user.

We extracted data from the following two key servers and created social network databases from them. The key rings were:

The Debian key-ring: The developers key-ring for the Debian distribution project was used as a small data set, using data captured as of July 5th, 2004. The key-ring as used by the GPG engine is about 10MB large.

The U Alberta key server key-ring: The key-ring of the U Alberta key server was used as a large data set, using data captured as of May 27th, 2005. The key-ring as used by the GPG engine is about 4GB large.

The Debian key-ring has about 1465 unique individuals within it, on average an individual has 1.6 keys. The U Alberta key server key-ring has about 830,000 unique individuals in it, each with an average of 2.38 keys. We hypothesise that this increase over the Debian data set is a result of the longitude of the key server data set. Within the Debian key-ring, there are about 17,912 keys which sign Debian maintainer's keys but are not part of the key-ring.

Furthermore, the email address enabled us to perform linking to other data sources. In the case of the Debian server, we were able to link the debian-devel and debian-project mailing list used by debian developers and extracted the social network information from it for comparison to the GPG key network. We matched email addresses to the individuals already linked to the GPG network and added new entries for people that were not.

4   General Network Properties

Figure 1 shows the growth of the number of individuals (nodes) and relationships (edges) in the social network incrementally built up using the Debian key-ring data.

Figure 2 shows the same growth for the social network created from the U Alberta key server data. We found that the large world of the U Alberta key server key-ring behaves in a manner similar to the bow-tie structure observed with the world wide web [11]. This bow-tie is composed of a core "knot" of relationships in the middle. This core is referred to by a large number of persons that are not in turn referred to (the left part of the bow-tie). The core also refers to a large number of people who do not refer back to the core (the right part of the bow-tie). Finally, there are a large number of "smaller worlds" unconnected to the rest of the key-ring (the lint).

A measure of the overall connectivity was computed for both data sets over time by picking a random individual and attempting to find a path to another random individual within the data set. By computing the number of successful paths between each pair, the overall connectivity for each key set was tracked.

Interestingly, as the size of the world grows, the likelihood that people will be in different connected components increases. The connectivity of the system within the key-ring was measured by tracking the number of times a path could be found from point A to point B. Overall the connectivity of the graph begins at 25% and drops to 3% when all individuals are within the world.

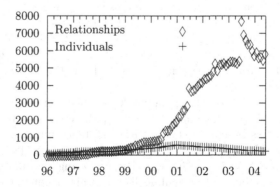

Fig. 1. Population and relationships within the Debian key-ring over time

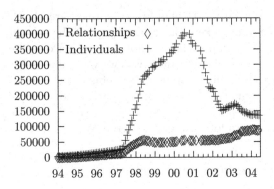

Fig. 2. Population and relationships within the U Alberta key server over time

The Debian key-ring has a well-curated database and it's interconnectivity tends to converge to about .33 (i.e., about 1/3 of the time, a path can be found to another individual within the key server). In contrast, the overall connectivity of the U Alberta key server keeps lowering itself to about .01 with the passing of time. We hypothesise that this is a direct result of the intended purpose of both data sets. The Debian key-ring is cleaned and maintained to support the Debian development process whereas the U Alberta key server is used as a database which is not trimmed. Old, obsolete or broken keys therefore accumulate and pollute the key server whereas extra and/or useless information is pruned from the Debian key-ring. A large part of the problem comes from the widespread utilisation of keys with no expiration information and which remain for an excessive amount of time.

Figure 1 shows a comparison of the number of nodes and the number of connections in the Debian network over time. Note that the number of connected components increases in the same manner as the number of nodes. This provides evidence that in certain social networks, the number of connected components continuously vary over time. With such networks, caution is required not only with the assumption that the underlying graph is connected but also with the assumption that there are a constant number of connected components.

As previously mentioned, we also made use of two main Debian mailing lists to compare against the GPG social network. Out of the 17,305 individuals that posted to the mailing list, only 806 were part of the GPG social network. There exist many explanations for this difference, which may include the curation process that occurs with the Debian keyring and one-time posts to the mailing list.

Figure 3 compares the average network distances for both GPG and email data-sets. Interestingly, the email data-set rapidly converges to an average distance of slightly less than 3. We found this consistence surprising as we expected a higher amount of one-off postings and individuals within the mailing lists and thus a higher variation in the metrics. By inspecting the mailing lists we discovered that a number of the mailing lists contain a number of long running discussions between 2 or 3 individuals within the mailing lists. This explains

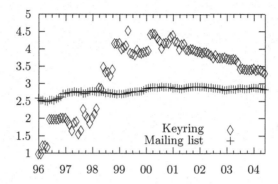

Fig. 3. Comparison of the distance in both mail and keyring social networks

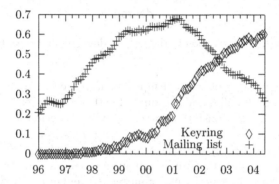

Fig. 4. Comparison of the connectivity in both mail and keyring social networks

the stability of the average social distance metric, however we are unsure of the reasons for the differences in the connectivity metrics between the GPG and mailing lists net that is plotted in Figure 4.

By inspecting the graph, it becomes obvious that the mailing lists lead the GPG key network temporally. This is what we expected intuitively as posting to a mailing list requires less preparation than creating a GPG key. Furthermore, the peak in mailing list connectivity also coincides with a number of Debian and Linux conferences already mentioned. We thus propose that the GPG network is a restricted subset of the mailing list network that lags behind because of its formalised structure.

5   Relationship Properties

For the Debian data set, the relationships follow a power law curve; on average each entity within the key-set would signal a relationship with about 3.8 other people (see Figure 5 for the degree distribution taken at an arbitrary time-step). Similarly, for the key-ring data set the relationships also follow a power-law curve

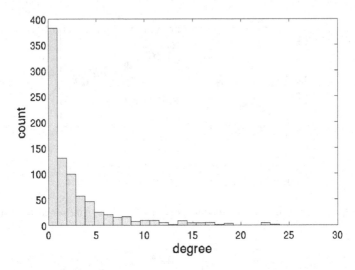

Fig. 5. Degree distribution in the Debian data set 4211 days from start

but the average number of relationships has decreased to 1.93. We believe that the number of single individuals accounts for this difference.

There are 15,939 relationships between individuals declared within the Debian key-ring. Out of these, 5,515 are symmetric in nature in that the relationship is reciprocated by the signee. The rest are one-way key signatures where an individual signs another key without any acknowledging signature. One possible reason for this behaviour is the use of automated email key signing methods. A review of the relationships did not, however, yield any obvious indicators of this.

These asymmetric relationships are analysed by Feld and Elmore [12] who suggested that they are present because of logistical difficulties in interacting with other persons or because individuals may select individuals which their peers consider popular but whom they themselves do not know personally. This may have some significance for managing cryptographic and trust networks, as it indicates that trust may be asymmetrical.

Within the U Alberta key server data set there are 118,960 distinct relationships declared, of which 69,193 are asymmetric. There are more than twice as many (2.8 times) directed relationships as there are symmetric relationships. Anecdotal evidence seems to support the proposal made by Feld and Elmore [12] that these specific asymmetric relationships are the result of social popularity and not actual acquaintance or social relationships. The most connected node within the key server data set is Phillip Zimmerman, the original author of the PGP^{TM} package. It is interesting that a 33% rule seems to be in effect—about 33% of all the relationships are asymmetric; this appears to be consistent with the results obtained from blogging data [13, 14].

Earlier, we argued that a popular metric for analysing social networks is the shortest path length between nodes in the network. Figure 6 demonstrates how the average shortest path length changes over time in both data sets.

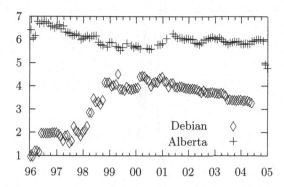

Fig. 6. Social distance change over time in both data sets

A typical use for this metric is for predicting which two nodes will be connected next in the development of the social network. When a new edge is added we are interested in the length of the previous shortest path between those two nodes (obviously after the addition of the new edge the length of the new shortest path will be one). If one can build a distribution over such lengths, it can be used to estimate the probabilities (for all possible pairs of nodes which are not already connected) that a particular edge will be added.

Figure 7 shows a kernel density function displaying over time the average shortest path between two nodes in the Debian data set before a relationship is added connecting them (the x-axis is time and the y axis is shortest path between nodes). In other words, at a time where there is a peak, when a new relationship is added between two nodes, the average shortest path between them is longer than when there is a valley. To put it another way, the peaks correspond to times when the people (nodes) in the network reach out farther in the graph for new relationships. Note that because we used a normalised kernel density function to display this data, y-axis has been rescaled. However, this representation clearly demonstrates the relative differences in the data over time.

Note the large peak in 1999 and the periodic peaks roughly every year there-after. We hypothesise that these peaks are explained by a number of Linux-based conferences—that contacts made while organising and attending the conference translated into key signatures. The conferences are as follows:

- Linux Expo 1995-1999, started 1995-06-26
- Linux World Expo, started 1999-08-09
- Linux Kongress 1994-2004, started 1995-05-01
- Linux Con Au 1999-2004, started 1999-07-09
- Linux Tag 1998-2004, started 1998-05-28
- Ottawa Linux Symposium 1999-2004, started 1999-07-22

The first edition of each conference is plotted in Figure 7 as a vertical black line and labelled. Subsequent editions are plotted as vertical grey lines. Clearly, the largest peak corresponds to the first edition of three of the conferences (Linux

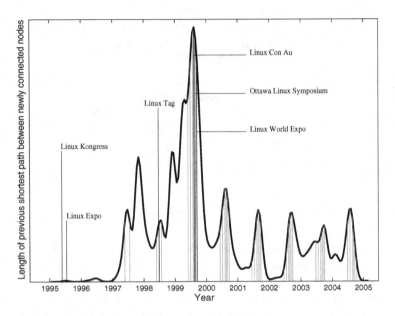

Fig. 7. Average shortest paths

Con Au, Ottawa Linux Symposium and Linux World Expo). Also, note that there is a high degree of correlation between the remaining conference dates and the peaks in the kernel density function.

This has important implications for both analysis and synthesis of social networks. If we gathered data from an existing hypothesised social network we could easily create such a graph of shortest paths over time. If there were distinct peaks in such a graph, it is reasonable to hypothesise that they correspond to events relevant to the social actors composing the network. This provides an useful research tool which narrows down a set of time periods within which researchers can search for such events. For example, there are some peaks in Figure 7 that do not correspond to the Linux conferences listed previously. This could be indicative of some other event of importance (similar to the conferences) to the Debian community. If one has some reason to believe that such an event exists these peaks could be useful in narrowing down the time-frame where the event could be found.

Alternately, if we wish to generate a "realistic" but synthetic social network modelling people's relationships through e-mail, we now have evidence that when determining which edges to add next as we build the graph, we should vary the probabilities of a possible edge over time to reflect the above behaviour. Perhaps by randomly generating times corresponding to important events where the probability of adding an edge between two more distant nodes should briefly spike as they do in Figure 7.

In the Debian data set the probability that a relationship joined two previously unconnected components of the graph is about 0.33. Figure 8 shows a kernel density function displaying this probability over time.

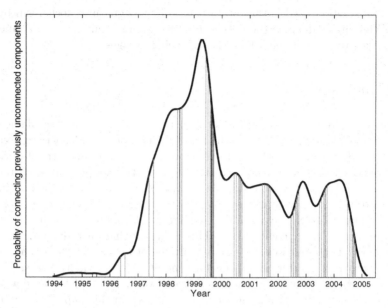

Fig. 8. Average number of edges that connect previously unconnected components

This behaviour in social networks can reasonably be interpreted in some sense as two separate groups making contact for the first time. This has important ramifications in such applications as the study of the spread of forest fires, or disease vectors.

Note that again there is some correlation between the largest mode in Figure 8 and the Linux conferences listed previously (the conferences are again plotted as vertical lines). More evidence that properties associated with social networks can vary significantly over time and thus should be tracked in a dynamic fashion.

Finally, an element of the GPG dataset that we found especially interesting is the insight into the privacy behaviour of individuals that it provides. As stated earlier, signatures between keys are required to ensure key authenticity and thus people tend to acquire signatures on an opportunistic basis for their own key.

As noted by Borisov et al. [15], this mechanism has privacy implications, indeed we have used it in this paper to acquire individuals' partial social networks. To an extent this constitutes a weakness of the system as it reveals a great deal of information to outside observers.

In the generic case, the individual makes use of his relationships to acquire signatures to solidify the authenticity of the public that he distributes. This ensures that a 'trusted' signature path exists between the sender's own key and the recipient's key, as the cost of exposing the social path between both persons.

An alternative, used to prevent information from being revealed, is to only sign ones own keys as they expire or get compromised. Provided that an alternate means of securing the distribution of the public key, this effectively prevents the release of social information to the key server.

A final solution used is the total dis-use of the signature mechanism by the user. While in a minority, these users tend to provide limited, cryptic labelling of their key to prevent the attribution of their messages.

6   Conclusion

We demonstrated how to build a social network using publicly available data from PGP^{TM} key servers—data which is ideal for the straightforward creation of a highly dynamic network. Next, we showed that two common small world related social network parameters, number of connected components and previous shortest path before a relationship, can change significantly over the life time of the network. Finally, we provided evidence that these changes can be related to events of interest to the actors in the social network (in the case of our data sets, the events were Linux conferences). This indicates that such dynamic analysis of these parameters could be useful in analysing other social networks as well as possibly providing better algorithms for generating synthetic networks.

There are two obvious directions for future work. The first is to see whether other social networks exhibit the behaviours shown in this paper. It is unlikely that the network of e-mail relationships is unique in this respect, but testing in other domains should be performed. The second is to use knowledge about the dynamic parameters to try and generate synthetic social nets and to see if these social nets are more "realistic" than those generated by static parameters. Specifically, we hypothesise that varying these parameters in a periodic manner will lead to an increase in the robustness of a generated network to disruption.

References

[1] Wilson, R.J.: Introduction to graph theory. John Wiley & Sons, Inc. (1986)
[2] Gross, J.L., Yellen, J., eds.: Handbook of graph theory. Discrete mathematics and its applications. CRC press (2004)
[3] Wasserman, S., Faust, K.: Social network analysis: methods and applications. Cambridge university press (1994)
[4] Carrington, P.J., Scott, J., Wasserman, S., eds.: Models and methods in social network analysis. Cambridge university press (2005)
[5] Watts, D.J., Strogatz, S.H.: Collective dynamics of 'small-world' networks. Nature **393** (1998) 440–442
[6] Milgram, S.: The small world problem. Psycology Today (1) (1967) 61–67
[7] Watts, D.: Small Worlds: The Dynamics of Networks between Order and Randomness. Princeton University Press (1999)
[8] Kleinberg, J.: The Small-World Phenomenon: An Algorithmic Perspective. In: Proceedings of the 32nd ACM Symposium on Theory of Computing. (2000)
[9] Hannerman, R.A.: Introduction to Social Network Methods. Department of Sociology, University of California (2001)
[10] Blaze, M., Feigenbaum, J., Lacy, J.: Decentralized trust management. In: IEEE Symposium on Security and Privacy. (1996) 164–173

[11] Broder, A., Kumar, R., Maghoul, F., Raghavan, P., Rajagopalan, S., Stata, R., Tomkins, A.: Graph structure in the web: Experiments and models. In: 9th World Wide Web Conference. (2000)

[12] Feld, S.L., Elmore, R.: Patterns of sociometric choices: Transitivity reconsidered. Social Psychology Quarterly **45**(2) (1982) 77–85

[13] MacKinnon, I., Warren, R.H.: Age and geographic analysis of the livejournal social network. Technical Report CS-2006-12, School of Computer Science, University of Waterloo (2006)

[14] Kumar, R., Novak, J., Raghavan, P., Tomkins, A.: Structure and evolution of blogspace. Commun. ACM **47**(12) (2004) 35–39

[15] Borisov, N., Goldberg, I., Brewer, E.: Off-the-record communication, or, why not to use pgp. In: Workshop on Privacy in the Electronic Society. (2004)

A Brief Survey of Machine Learning Methods for Classification in Networked Data and an Application to Suspicion Scoring

Sofus Attila Macskassy[1] and Foster Provost[2]

[1] Fetch Technologies,
2041 Rosecrans Ave, Suite 245, El Segundo, CA 90245
sofmac@fetch.com
[2] New York University,
Stern School of Business, 44 W. 4th Street, New York, NY 10012
fprovost@stern.nyu.edu

Abstract. This paper surveys work from the field of machine learning on the problem of within-network learning and inference. To give motivation and context to the rest of the survey, we start by presenting some (published) applications of within-network inference. After a brief formulation of this problem and a discussion of probabilistic inference in arbitrary networks, we survey machine learning work applied to networked data, along with some important predecessors—mostly from the statistics and pattern recognition literature. We then describe an application of within-network inference in the domain of suspicion scoring in social networks. We close the paper with pointers to toolkits and benchmark data sets used in machine learning research on classification in network data. We hope that such a survey will be a useful resource to workshop participants, and perhaps will be complemented by others.

1 Introduction

This paper briefly surveys work from the field of machine learning, summarizes work in a trio of research papers [1,2,3]. This extended abstract consists of the abstracts for those papers in which we concentrate on methods published in the machine learning literature, as well as methods from other fields that have had considerable impact on the machine learning literature.

Networked data are the special case of relational data where entities are interconnected, such as web-pages or research papers (connected through citations). We focus on *within-network* inference, for which training entities are connected directly to entities whose classifications (*labels*) are to be estimated. This is in contrast to *across-network* inference: learning from one network and applying the learned models to a separate, presumably similar network [4,5]. For within-network inference, networked data have several unique characteristics that both complicate and provide leverage to learning and inference.

E.M. Airoldi et al. (Eds.): ICML 2006 Ws, LNCS 4503, pp. 172–175, 2007.

Although the network may contain disconnected components, generally there is not a clean separation between the entities for which class membership is known and the entities for which estimations of class membership are to be made. The data are patently not i.i.d., which introduces bias to learning and inference procedures [6]. The usual careful separation of data into training and test sets is difficult, and more importantly, thinking in terms of separating training and test sets obscures an important facet of the data. Entities with known classifications can serve two roles. They act first as training data and subsequently as background knowledge during inference. Relatedly, within-network inference allows models to use specific node identifiers to aid inference [7].

Network data generally allow *collective inference*, meaning that various interrelated values can be inferred simultaneously. For example, inference in Markov random fields [8] uses estimates of a node's neighbor's labels to influence the estimation of the nodes labels—and vice versa. Within-network inference complicates such procedures by pinning certain values, but again also offers opportunities such as the application of network-flow algorithms to inference. More generally, network data allow the use of the features of a node's neighbors, although that must be done with care to avoid greatly increasing estimation variance (and thereby error) [9].

2   Network Learning

Abstract from [1]:
This paper presents NetKit, a modular toolkit for classification in networked data, and a case-study of its application to networked data used in prior machine learning research. We consider *within-network classification*: entities whose classes are to be estimated are linked to entities for which the class is known. NetKit is based on a node-centric framework in which classifiers comprise a local classifier, a relational classifier, and a collective inference procedure. Various existing node-centric relational learning algorithms can be instantiated with appropriate choices for these components, and new combinations of components realize new algorithms. The case study focuses on univariate network classification, for which the only information used is the structure of class linkage in the network (i.e., only links and some class labels). To our knowledge, no work previously has evaluated systematically the power of class-linkage alone for classification in machine learning benchmark data sets. The results demonstrate that very simple network-classification models perform quite well—well enough that they should be used regularly as baseline classifiers for studies of learning with networked data. The simplest method (which performs remarkably well) highlights the close correspondence between several existing methods introduced for different purposes—i.e., Gaussian-field classifiers, Hopfield networks, and relational-neighbor classifiers. The results also show that a small number of component combinations excel. In particular, there are two sets of techniques that are preferable in different situations, namely when few versus many labels

are known initially. We also demonstrate that link selection plays an important role similar to traditional feature selection.

3   Suspicion Scoring

Abstract from [2]:
We describe a guilt-by-association system that can be used to rank entities by their suspiciousness. We demonstrate the algorithm on a suite of data sets generated by a terrorist-world simulator developed under a DoD program. The data sets consist of thousands of people and some known links between them. We show that the system ranks truly mali-cious individuals highly, even if only relatively few are known to be malicious ex ante. When used as a tool for identifying promising data-gathering opportunities, the sys-tem focuses on gathering more information about the most suspicious people and thereby increases the density of link-age in appropriate parts of the network. We assess per-formance under conditions of noisy prior knowledge (score quality varies by data set under moderate noise), and whether augmenting the network with prior scores based on profiling information improves the scoring (it doesn't). Although the level of performance reported here would not support direct action on all data sets, it does recommend the consideration of network-scoring techniques as a new source of evidence in decision making. For example, the system can operate on networks far larger and more com-plex than could be processed by a human analyst.

Abstract from [3]:
We describe a guilt-by-association system that can be used to rank networked entities by their suspiciousness. We demonstrate the algorithm on a suite of data sets generated by a terrorist-world simulator developed to support a DoD program. Each data set consists of thousands of entities and some known links between them. The system ranks truly malicious entities highly, even if only relatively few are known to be malicious ex ante. When used as a tool for identifying promising data-gathering opportunities, the system focuses on gathering more information about the most suspicious entities and thereby increases the density of linkage in appropriate parts of the network. We assess performance under conditions of noisy prior knowledge of maliciousness. Although the levels of performance reported here would not support direct action on all data sets, the results do recommend the consideration of network-scoring techniques as a new source of evidence for decision making. For example, the system can operate on networks far larger and more complex than could be processed by a human analyst. This is a follow-up study to a prior paper; although there is a considerable amount of overlap, here we focus on more data sets and improve the evaluation by identifying entities with high scores simply as an artifact of the data acquisition process.

References

1. Macskassy, S.A., Provost, F.: Classification in Networked Data: A toolkit and a univariate case study. Technical Report CeDER Working Paper 04-08, Stern School of Business, New York University (2004). [June 2006 revision]
2. Macskassy, S.A., Provost, F.: Suspicion scoring based on guilt-by-association, collective inference, and focused data access. In: International Conference on Intelligence Analysis. (2005)
3. Macskassy, S.A., Provost, F.: Suspicion scoring of entities based on guilt-by-association, collective inference, and focused data access. In: Annual Conference of the North American Association for Computational Social and Organizational Science (NAACSOS). (2005)
4. Craven, M., Freitag, D., McCallum, A., Mitchell, T., Nigam, K., Quek, C.Y.: Learning to Extract Symbolic Knowledge from the World Wide Web. In: 15th Conference of the American Association for Artificial Intelligence. (1998)
5. Lu, Q., Getoor, L.: Link-Based Classification. In: Proceedings of the 20th International Conference on Machine Learning (ICML). (2003)
6. Jensen, D., Neville, J.: Linkage and Autocorrelation Cause Feature Selection Bias in Relational Learning. In: Proceedings of the 19th International Conference on Machine Learning (ICML). (2002)
7. Perlich, C., Provost, F.: Distribution-based aggregation for relational learning with identifier attributes. Machine Learning $62(1/2)$ (2006) 65–105
8. Besag, J.: Spatial interaction and the statistical analysis of lattice systems. Journal of the Royal Statistical Society $36(2)$ (1974) 192–236
9. Jensen, D., Neville, J., Gallagher, B.: Why Collective Inference Improves Relational Classification. In: Proceedings of the 10th ACM SIGKDD International Conference on Knowledge Discovery and Data Mining. (2004)

Age and Geographic Inferences of the LiveJournal Social Network

Ian MacKinnon and Robert Warren

David R. Cheriton School of Computer Science,
University of Waterloo, Waterloo, Ontario, Canada
{imackinn,rhwarren}@uwaterloo.ca

Abstract. Online social networks are often a by-product of blogging and other online media sites on the Internet. Services such as LiveJournal allow their users to specify who their "friends" are, and thus a social network is formed. Some users choose not to disclose personal information which their friends list. This paper will explore the relationship between users with the intent of being able to make a prediction of a users age and country of residence based on the information given by their friends on this social network.

1 Introduction

We review the preliminary results of our analysis of a partial LiveJournal data set. The findings here represent the first stages of a larger project to analyze the people and relationships that bind them online. We consider the linking between the global location of users in LiveJournal with countries of their friends. We also intend to look at the relationship between the age of a user and the age of their friends. The obvious application of knowing how strong this bond is would be to infer demographic information based on data we know about their friends[1].

2 Data Collection

Initial user discovery was accomplished by polling the LiveJournal "Latest Posts"[1] feed for the first week of September 2005. Of users discovered, using only those who identified themselves to be from the top 8 countries as a start, a breadth-first search was performed by crawling the users information page on LiveJournal. For each user crawled, any demographic data volunteered by the user was recorded. This process was performed at small intervals over the course of September to December 2005. In total, information from 4,138,834 LiveJournal users were collected. A subset of 2,317,517 users entered our sample, who had self-reported to be from one of the top 8 nations. Table 1 shows the number of users from each of the top 8 countries. As identified by [2], there are 2 limiting factors when dealing with self-reported data: Many users do not report their location or age or report erroneous or false data.

---

[1] http://www.livejournal.com/stats/latest-rss.bml

E.M. Airoldi et al. (Eds.): ICML 2006 Ws, LNCS 4503, pp. 176–178, 2007.
© Springer-Verlag Berlin Heidelberg 2007

3   Is Age a Factor in Relationships?

We gathered age data for all of the users within the data set (effective Dec. 31, 2005) and calculated the average age of the friends for each individual user. We also gathered the standard deviation for each group of friends and clustered the information according to users age. We attempted a linear regression classifier based on the normal age mean of the immediate social circle, divided by the user's age. We randomly split the data in two equal sets and calculated the mean age of their social network. Using the first set, we calculated the average slope of the user age versus the mean social network age, which is 0.992. We then used the second set to benchmark the precision of the classifier at different prediction interval.

Table 2 represents the different precisions obtained within certain confidence intervals. Through experimentation with linear regressions and other classifications methods, we have concluded that there does exist a relationship between the age of a person and their peer group.

4   Inferring User Location

We considered a lookup table of probabilities, where we can lookup what percentage of a users friends are from a country, X, and see what the probability is that a user is also from country X. For example, if we know that 30% of a user's friends are Canadian, what is the probability that the user is Canadian himself? Given our new subset of information, we iterate through all the users and for each country and divide the number of friends that user has from that country by their total number of friends.

From this, we can generate Fig 1, which establishes a general trend in regards to the relationship between people with these percentage of friends from a country, and the probability they are also from that country. Interestingly, we see that Americans and Russians have a radically different curve from the rest.

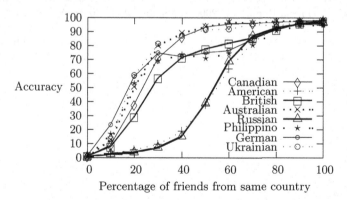

Fig. 1. Probability of country of residence based on the dominant country of friends

Table 1. Top 8 countries of origin reported by users on LiveJournal

Country	Count
United States	2990918
Russian Federation	252258
Canada	233839
United Kingdom	191650
Australia	89729
Philippines	31230
Germany	29224
Ukraine	28478

Table 2. Predict user's age based on the mean friend age

Age Range(+/-)	Precision
6 months	0.29
8 months	0.39
1 year	0.49
1 1/2 years	0.62
2 years	0.71
2 1/2 years	0.76
3 years	0.80
3 1/2 years	0.83
4 years	0.86
5 years	0.98

This indicates that many people who are not American or Russian, have a large number of American or Russian friends. We found that the more friends a user has, the less accurate our ability to predict their country is. This result seems to indicate that as a user gets more friends, they get more and more from outside their home country.

5   Future Work

We plan to extend the age and location models presented here to not only find falsely reported location and age information, but also correct it. Also, we wish to see if results are similar on other online social networks.

References

1. MacKinnon, I. and Warren, R.H.: Age and geographic analysis of the livejournal social network. Technical Report CS-2006-12, School of Computer Science, University of Waterloo, April 2006.
2. Lin, J., Halavais, A.: Mapping the blogosphere in America. Workshop on the Weblogging Ecosystem at the 13th World Wide Web Conference (WWW2004), New York City, USA (2004)

Inferring Organizational Titles in Online Communication

Galileo Mark S. Namata Jr.[1], Lise Getoor[1], and Christopher P. Diehl[2]

[1] Dept. of Computer Science/UMIACS,
Univ. of Maryland, College Park, MD 20742
[2] Johns Hopkins Applied Physics Lab.,
11100 Johns Hopkins Rd., Laurel, MD 20723

1 Introduction

There is increasing interest in the storage, retrieval, and analysis of email communications. One active area of research focuses on the inference of properties of the underlying social network giving rise to the email communications[1,2]. Email communication between individuals implies some type of relationship, whether it is formal, such as a manager-employee relationship, or informal, such as friendship relationships. Understanding the nature of these observed relationships can be problematic given there is a shared context among the individuals that isn't necessarily communicated. This provides a challenge for analysts that wish to explore and understand email archives for legal or historical research.

In this abstract, we focus on a specific subproblem of identifying the hierarchy of a social network in an email archive. In particular, we focus on and define the problem of inferring a formal reflection of an organizational hierarchy, the formal title of an individual, within the underlying social network. We present a new dataset, to use in conjunction with the original Enron dataset, for studying the formal organizational structure underlying an email archive. We also provide preliminary results from the classification of individuals to broad titles in the organization, relying only on simple traffic statistics.

2 Enron Hierarchy Dataset

One major impediment to research in hierarchy inference from email archives is the lack of a publicly available dataset providing email traffic from a structured organization along with documentation of that structure. Therefore, we present a dataset[1] for use with the Enron email dataset[3], with a particular focus on identifying the identities and titles of the individuals from whose accounts the Enron email dataset was generated. Using various forms of the dataset[4,5,6], along with documents related to the Enron trial, we identified the individuals whose accounts compose the Enron dataset, as well as the email addresses, titles and company groups for 124 of them. We also provide a mapping to six broad titles from similar titles i.e.: VP (Vice President) for VP of Finance.

---

[1] Available from http://www.cs.umd.edu/projects/linqs.

E.M. Airoldi et al. (Eds.): ICML 2006 Ws, LNCS 4503, pp. 179–181, 2007.
© Springer-Verlag Berlin Heidelberg 2007

3   Problem

In order to identify the underlying social network hierarchy, we focus on a formal representation of this hierarchy, the formal titles of individuals. Specifically, we focus on the problem of mapping the set of actors (people who send/receive emails in an email collection) to a set of formal titles within the organization.

In our experiments, we classify the 124 individuals in our extended dataset to six broad titles using various classifiers[7] processing simple statistics derived from the relevant traffic. We use three types of traffic statistics: undirected (# of emails sent and received), directed(# of emails sent, # of emails received) and aggregate (# of emails sent/received to labeled individuals of a given broad title). The goal is to identify which set of statistics yields the best performance.

Fig. 1 shows a summary of the results from different classifiers using our various traffic models. Given the class distribution of the six target broad titles has an average random classifier performance of 20% accuracy, the results are promising. In general, our classifiers outperformed the random baseline by a statistically significant margin. Using our undirected model, we received an accuracy of 53.2%, over twice as well as random. We note that although the undirected model lacks traffic direction information, it performed comparably with the directed model. The same is true in variations where we only used one direction of the traffic. This implies that we may be able to classify individuals without having all their email communications. Finally, we note that when we use our aggregate model with the other two models, we are able to reach an accuracy of over 62%.

It is also interesting to examine the confusion matrix, in Fig. 2, to see where the misclassifications are occurring. Of note is the correlation between the misclassifications of individuals to titles. The misclassification of a title seems to occur mainly with titles close to the correct title in the hierarchy. For example, *DIR* is mainly misclassified with its immediate superior, *VP*, and subordinate, *MGR*. Similarly, *ASSOC* is most misclassified as *MGR*. This trend is consistent

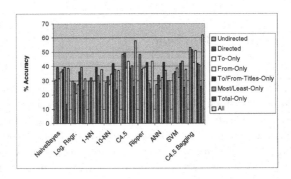

Fig. 1. Summary of Broad Title Classification Accuracy

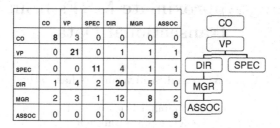

	CO	VP	SPEC	DIR	MGR	ASSOC
CO	8	3	0	0	0	0
VP	0	21	0	1	1	1
SPEC	0	0	11	4	1	1
DIR	1	4	2	20	5	0
MGR	2	3	1	12	8	2
ASSOC	0	0	0	0	3	9

Fig. 2. Confusion matrix for broad titles correspond to title hierarchy

among all the titles. This implies that our approach using traffic statistics might be able to reconstruct levels in the overall hierarchy.

4 Conclusion

Identifying the hierarchy of the underlying social network is an important problem which aids in the exploration and understanding of organizational email archives. We investigate a component of this larger problem, mapping individuals to their formal titles in the organization, using only simple traffic statistics and present preliminary experimental results. In future work, we would like to make greater use of commonly used social network analysis measures such as centrality and equivalence in our classifiers, as well as use the content of the email messages. We are also interested in trying to classify other relationships, such as friendships or direct report relationships. Finally, we are interested in the temporal aspects of the hierarchy, specifically being able to detect how and when the organizational structure changes in an archive.

References

1. Diehl, C., Getoor, L., Namata, G.: (Name reference resolution in organizational email archives) 6th SIAM Conference on Data Mining.
2. Corrada-Emmanuel, A., McCallum, A., Wang, X.: Topic and role discovery in social networks. In: IJCAI. (2005)
3. Klimt, B., Yang, Y.: Introducing the Enron corpus. In: Conference on Email and Anti-Spam. (2004)
4. Adibi, J.: Enron dataset (2005)
5. Fiore, A.: UC Berkeley Enron email analysis (2005) http://bailando.sims. berkeley.edu/enron_email.html.
6. Corrada-Emmanuel, A.: Enron email dataset research (2004) http:// ciir.cs.umass.edu/~corrada/enron.
7. Witten, I., Frank, E.: Data Mining: Practical machine learning tools and techniques. Volume 2nd Edition. Morgan Kaufmann (2005)

Learning Approximate MRFs from Large Transactional Data*

Chao Wang and Srinivasan Parthasarathy

Department of Computer Science and Engineering,
The Ohio State University

1 Introduction

In this abstract we address the problem of learning approximate *Markov Random Fields* (MRF) from large transactional data. Examples of such data include market basket data, co-authorship networked data, etc. Such data can be represented by a binary data matrix, with an entry (i, j) takes a value of one (zero) if the item j is (not) in the basket i. "Large" means that there can be many rows or columns in the data matrix. To model such data effectively in order to answer queries about the data efficiently, we consider the use of probabilistic models. In this abstract, we consider employing frequent itemsets to learn approximate global MRFs on large transactional data. We conduct an empirical study on real datasets to show the efficiency and effectiveness of our model on solving the query selectivity estimation problem, that is to approximately compute the marginal probability of sets of items (see [1] for the experimental results). Translated into the social network domain, this is the problem of computing the likelihood of seeing a particular combination of grocery items in the market basket domain, or the probability of a group of professors coauthoring a paper in a co-authorship network, etc. This marginal probability computation is also useful for anomalous link detection [2] in social network analysis. A link in a social network corresponds to a pair of items. The links whose associated marginal probabilities are significantly low can be thought of as anomalous.

2 Background

Let $\mathcal{I}$ be a set of items, i_1, i_2, ..., i_d. A subset of $\mathcal{I}$ is called an *itemset*. The *size* of an itemset is the number of items it contains. An itemset of size k is a k-itemset. Translated into the social network domain, an item corresponds to an actor and an itemset corresponds to a group of actors in a network. Take market basket data as an example, items here correspond to grocery items, and itemsets correspond to the contents of individual baskets. A transactional dataset is a collection of itemsets, $D = \{t_1, t_2, \ldots, t_n\}$, where $t_i \subseteq \mathcal{I}$. For any itemset α,

---

* This work is supported by DOE Award No. DE-FG02-04ER25611 and NSF CAREER Grant IIS-0347662. We refer the reader to a longer version of this paper [1] for experimental results and complete proofs and discussions.

E.M. Airoldi et al. (Eds.): ICML 2006 Ws, LNCS 4503, pp. 182–185, 2007.

we write the transactions that contain α as $D_\alpha = \{t_i | \alpha \subseteq t_i \text{ and } t_i \in D\}$. Each item is modeled as a random variable.

Definition 1. *(Frequent itemset): For a transactional dataset D, an itemset α is frequent if $|D_\alpha| \geq \sigma$, where $|D_\alpha|$ is called the support of α in D, and σ is a user-specified non-negative threshold.*

Using Frequent Itemsets to Learn an MRF: The idea of using frequent itemsets to learn an MRF was first proposed by Pavlov *et al.* [3]. A k-itemset and its support represents a k-way statistic and can be viewed as a constraint on the true underlying distribution that generates the data. Given a set of itemset constraints, a *Maximum Entropy* (ME) distribution satisfying all these constraints is selected as the estimate for the true underlying distribution. This ME distribution is essentially equivalent to an MRF. A simple iterative scaling algorithm can be used to learn an MRF from a set of itemsets and efficient inference is crucial to the running time of the learning algorithm. We call those models learned through exact inference procedures *exact* models. The *junction tree* algorithm is a commonly-used exact inference engine for probabilistic models. The time complexity of the algorithm is exponential in the *treewidth* of the underlying model. For real-world models, it is quite common that the treewidth will be well above 20, making learning exact models intractable. As a result, we have to resort to learning approximate models. Pavlov *et al.* [3] did not solve the problem of learning MRFs on large transactional data. The MRFs they generate target specifically the query variables and are therefore quite lightweight. In another related work, Goldenberg and Moore [4] proposed to use frequent itemsets to learn Bayesian networks over all variables.

3   Learning Approximate MRFs

Let us first consider an extreme case in which the whole graphical model consists of a set of disjoint non-correlated components. Then the joint distribution can be obtained in a straightforward fashion: Given an undirected graphical model G subdivided into disjoint components C_1, C_2, ..., C_n (not necessarily connected components), the probability distribution associated with G is given by: $p(X) = \prod_{i=1}^{n} p(X_{C_i})$.

3.1   Clustering Variables Based on Graph Partitioning

The basic idea of our proposed divide-and-conquer style approach comes directly from the above observation. Specifically, the variables are clustered into groups according to their correlation strengths. We call such a group a *variable-cluster*. Then a local MRF is inferred on each *variable-cluster*. In the end we aggregate all the local models to obtain a global model. k-MinCut [5] can serve our purpose of clustering variables. Each graph partition corresponds to a *variable-cluster*. Intuitively, we want to maximize correlations among variables within *variable-clusters*, and minimize correlations among variables across *variable-clusters*. To accomplish this we ensure that the weight of edges reflect the strength of correlations between variables. To this end, we propose a cumulative weighting scheme as follows: for each itemset of size ≥ 2 , we add its support to the weight of all related edges, whose two vertices are contained by the itemset.

3.2 Interaction Importance and Treewidth Based Variable-Cluster Augmentation

The *variable-clusters* produced by the k-MinCut partitioning scheme are disjoint. Intuitively, there can be correlation information that is lost during the partitioning. To compensate for this loss, we propose an interaction importance based *variable-cluster* augmenting scheme. The idea is that we allow each *variable-cluster* to grow outward. More specifically, it attracts and absorbs most important interactions (edges) incident to its vertices from outside to itself. As a result, some extra variables are pulled into the *variable-cluster*. We control the augmentation through the number of extra vertices pulled into the cluster (called *growth factor*). One can use the same growth factor for all *variable-clusters* to preserve their balance.

As an optimization, we account for the model complexity during the augmentation. We keep augmenting a partition until its complexity reaches a user-specified threshold. More specifically, we keep track of the growth of the treewidth during the augmentation. 1-hop neighboring vertices are first considered for the augmentation, followed by 2-hop neighboring vertices and so on. Meanwhile, we still follow the interaction importance criteria. The resultant augmented partitions are likely to become unbalanced in terms of their size. The partitions with a small treewidth will grow more significantly than those with a large treewidth. However, these partitions are balanced in terms of their complexity. A benefit is that more interactions across partitions will be accounted for in a computationally controllable manner, leading to a more accurate global model.

3.3 Approximate Global MRFs and a Greedy Inference Algorithm

For each augmented *variable-cluster*, we collect all of its related itemsets and use the iterative scaling algorithm to learn an exact local model. Two local models are correlated to each other if they share variables. The collection of all local models forms a global model of the original data. How do we make inferences on this model efficiently? If all query variables are subsumed by a single local MRF, we just need to calculate the marginal probability within that model. If they span multiple local models, we use a greedy decomposition scheme. First, we pick the local model that has the largest intersection with the current query (i.e., covers most query variables). Then we pick the next local model that covers most uncovered query variables. This covering process will be repeated until we cover all query variables. Simultaneously, all intersections between the above local models and the query are recorded. In the end, we derive an overlapped decomposition of the query and we use Lemma 1 to compute its marginal probability.

Lemma 1. *Given an undirected graphical model G subdivided into n overlapped components, if there exists an enumeration of these n components, i.e., C_1, C_2, ..., C_n, s.t., for any $2 \leq i \leq n$, the separating set, $s(C_i, \cup_{j=1}^{i-1} C_j) \subseteq (C_i \cap (\cup_{j=1}^{i-1} C_j))$, then the probability distribution associated with G is given by:* $p(X) = \dfrac{\Pi_{i=1}^{n} p(X_{C_i})}{\Pi_{i=2}^{n} p(X_{C_i} \cap (\cup_{j=1}^{i-1} X_{C_j}))}$.

The greedy inference scheme is a heuristic, since it is possible to have cyclic dependencies among the decomposed pieces. Also, our global model is not strictly globally consistent in that there can exist inconsistencies across the local models.

4   Conclusion

In this abstract, we have described a new approach to learning approximate MRFs on large transactional data. Our proposed approach has been shown to be very effective and efficient in solving the selectivity estimation problem. In the future, we would like to exploit the learned models on various social network analysis tasks, such as link prediction and anomalous link detection.

References

1. Wang, C., Parthasarathy, S.: Learning approximate MRFs from large transactional data. In: The Ohio State University, Technical Report: OSU-CISRC-5/06–TR59. (2006)
2. Rattigan, M.J., Jensen, D.: The case for anomalous link discovery. ACM SIGKDD Explorations Newsletter **7** (2005) 41–47
3. Pavlov, D., Mannila, H., Smyth, P.: Beyond independence: probabilistic models for query approximation on binary transaction data. IEEE Transactions on Knowledge and Data Engineering **15** (2003) 1409–1421
4. Goldenberg, A., Moore, A.: Tractable learning of large Bayes net structures from sparse data. In: Proceedings of the twenty-first international conference on Machine learning. (2004)
5. Karypis, G., Kumar, V.: Multilevel k-way partitioning scheme for irregular graphs. J. Parallel Distrib. Comput. **48** (1998) 96–129

Panel Discussion

David M. Blei

Princeton University
Princeton, NJ 08544, USA
blei@cs.princeton.edu

In this volume, we have seen several compelling reasons for the statistical analysis of network data.

1. Find statistical regularities in an observed set of relationships between objects. For example, what kinds of patterns are there in the friendships between co-workers?
2. Understand and make predictions about the specific behavior of certain actors in a domain. For example, who is Jane likely to be friends with given the friendships we know about?
3. Make predictions about a *new* actor, having observed other actors and their relationships. For example, when someone new moves to town, what can we predict about his or her relationships to others?
4. Use network data to make predictions about an actor-specific variable. For example, can we predict the functions of a set of proteins given only the protein-protein interaction data?

All of the analysis techniques proposed here are model-based: one defines an underlying joint probability distribution on graphs and considers the observed relationship data under that distribution. Loosely—and this will be a point of discussion among the panelists—the models can be divided into those that are "descriptive" or "discriminative" and those that are "generative."

A descriptive graph model is one where the number of nodes in the observed graph or graphs is held fixed and the joint distribution is defined over the edges of that fixed set. The influential exponential random graph model is a general formulation of a descriptive graph model [1,2]. In this framework, the distribution of the entire graph structure is an exponential family with sufficient statistics that are aggregates of the entire graph, e.g., the number of triangles.

In a generative graph model, there is a clear probabilistic mechanism for expanding the graph to new nodes and new edges. The paper by Goldenberg and Zheng is a full generative graph model: the joint distribution is built around the notion of new actors and new connections between existing actors. There is still a joint distribution over the observed graphs. However, the probability of a new node is well-defined and the probability of a new edge can be computed without recalibrating the distribution.

There is ample room for middle ground between these categories. Several papers define hierarchical models based on the latent space approach [3]. These models are generative in the sense that new edges are conditionally independent of the others and have a well-defined probabilistic generative process. But they

E.M. Airoldi et al. (Eds.): ICML 2006 Ws, LNCS 4503, pp. 186–194, 2007.
© Springer-Verlag Berlin Heidelberg 2007

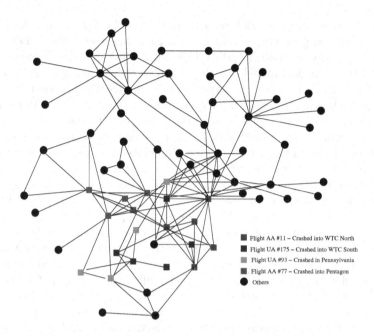

Fig. 1. 9-11 Hijacker/Terrorist Network. Source: [4].

are somehow not "as generative" as Goldenberg and Zheng's model, where the evolution of the social network is part of the fabric of the generative process.

This distinction was only one of the issues addressed in the workshop that accompanied this volume. As in any kind of data analysis, the tools required depend on the job at hand. We saw work on modeling sequential observations from social networks, modeling multiple data types such as citations and text, fitting graphs organized into hierarchies, and developing new statistics for the exponential random graph model.

Many of these tools were presented for the first time at the workshop. In this panel discussion, we have asked some of our distinguished participants to reflect on the contents and offer a comparative perspective.

Stephen E. Fienberg
Carnegie Mellon University
Pittsburgh, PA 15213, USA
fienberg@stat.cmu.edu

The papers at this workshop when taken together capture many fascinating aspects of network modeling. Prodded by some earlier discussion, I thought it would be useful to begin by reminding us about the two different kinds of graphical representations of the traditional $n \times p$ data array for n individuals or units by p variables. Graphical models [5] are used to represent relationships among the variables in terms of independence relationships. Graphical representations for networks are used to represent relationships among the units, and it is precisely because we don't have conditional independences in the usual dyadic models that things are somewhat complex. Our goal in this workshop has been to both focus on the latter kind of network models and to show how to link them in different ways to probabilistic/statistical models for the variables, either through representations of covariate information (see the paper here by Handcock) or or via mixed membership models (see the papers here by Airoldi et al. and by McCallum et al.).

As others have already noted, we have seen two different types of network models—generative (or what Shalizi et al. call agent-based in their paper) and descriptive. In the latter, which includes the class of p^* models described here by Wasserman et al., we identify *motifs* such as triads or stars and then build models that use them as primary data summaries, e.g., sufficient statistics. When we focus on the evolution of networks we can often be blending the two types of models although they can still be purely descriptive, c.f. the paper here by Henneke and Xing. An interesting question we have raised is whether the latent space models of [3] are descriptive or generative. At one level they appear to be descriptive but they are quite similar in other ways to the mixed membership stochastic block models in the papers by Airoldi et al. and by McCallum et al., which are generative in nature see also [6] and the discussion that follows it.

One issue on which we have not dwelled but which is implicit in the discussions of the distinction between the models is the nature of the data at hand. When we ask what are the data and where do they come from we are really asking a generative question which frames the nature of the models we should be considering. Consider the reported "network" demonstrating the links among the 9/11 hijackers that the press and administration officials are so fond of describing. Figure 1 shows perhaps the most carefully constructed version of it due to [4] What types of linkages do the edges in the graph represent, i.e., to what variables do they correspond? Was the graph constructed to assure that there are paths linking the hijackers to one another? The network picture shows linkages to others beyond the 9/11 hijackers with Arabic names. Are these individuals to be considered hijacker accomplices or confederates? What about others to who linkages could have been made beyond the horizon of observability from each other? After all, many hijackers on the same flight were more than 2 steps away from each other. Finally, real linkages in a terrorist network are dynamic but Figure 1 represents data collapsed over time.

Let me end by summarizing what I see as three major statistical modeling challenges in the analysis of network data. These relate to both the quality and the ease of inference:

Computability. Can we do computations exactly for large networks, e.g., by full MCMC methods, or do we need to resort to approximations such as those involved in the variational approximation such as in the paper by Airoldi et al.?

Asymptotics. The is no standard asymptotics for networks, e.g., as n goes to infinity, which can be used to assess goodness of fit of models. Thus we may have serious problems with variance estimates for parameters and with confidence or posterior interval estimates. The problem here is the inherent dependence of network data.

Embeddibility. Do our data represent the entire network or are they based on only a subnetwork or subgraph, as in Figure 1? When the data come from a subgraph we need to worry about boundary effects and the the attendant bias they bring to parameter estimates. The one result I know in this area is due to [7] for scale-free models in which they show the extent and nature of the bias. My suspicion is that there are similar issues for most of the models discussed at this workshop and we need to explore the consequences of these.

Andrew McCallum
University of Massachusetts
Amherst, MA 01003
mccallum@cs.umass.edu

Task-Focussed Social Network Analysis

I am a relative newcomer to social network analysis. Although I have been doing some research in SNA for the past few years, most of my research over the past decade has been in natural language processing. With this "outsider's perspective," I'd like to offer a couple of thoughts about possible fruitful future directions for SNA.

First, I encourage work in discriminatively-trained social network models.

Many of the recently-proposed models in my local sub-area of SNA are generative directed graphical models. These include various mixed-membership "topic models" and related models, such as *author-topic* [8], *author-recipient-topic* [9], *role-author-recipient-topic* [10], *group-topic* [11], *infinite-relational* [12], *entity-topic* [13], *relational mixed membership* [14], and *community-user-topic* [15]. Other generative models are mentioned by the other panelists.

Although research NLP was dominated by generative models (such as hidden Markov models and probabilistic context free grammars) for decades, the past five years have seen a much stronger emphasis on "discriminative" conditional-probability-trained models, such as logistic regression, maximum entropy models and conditional random fields. Here model parameters are estimated by maximizing the conditional probability of some output variables given some input context. Because the model is not responsible for generating the input context, we need not be concerned about violated independence assumptions among the input variables, and we are free to use rich, expressive sets of overlapping input features. In NLP, the move from generative models to discriminative models typically yields significant gains in accuracy.

Like in natural language, social network data sets are often rich in context, multiple modalities, and other non-independent variables that would benefit from a discriminative approach. We have begun research toward "conditionally-trained topic models" with our work on *multi-conditional learning*, and in particular *multi-conditional mixture models* [16].

Second, in a related point, I encourage emphasis on particular tasks.

Much past work in SNA approaches the problem as a scientist—we observe some natural phenomenon, and attempt to build models that capture them.

These include foundational work in descriptive graph properties, generative models of graphs, etc. It is also interesting (and sometimes more useful) to approach a domain as an engineer—asking "What is the real task we are trying to solve?" "What is the use-case?" "What is the decision problem?"

There are, of course, many important use-cases for social network analysis: deciding who to promote, finding an expert, selecting the right actions to improve (or harm) an organization, identifying likely illicit behavior, selecting the best collaborator, finding new music I'm likely to enjoy, predicting which team will get the job done best.

Scientifically descriptive models may have something to say about these tasks, but discriminative SNA models could focus on tuning their parameters for best accuracy on these particular tasks. As interest in SNA expands, I predict that there will be more research on models designed to address particular tasks.

Cosma Rohilla Shalizi
Carnegie Mellon University
Pittsburgh, PA 15213, USA
cshalizi@cmu.edu

Looking back over the papers presented at this workshop, I am struck by two cross-cutting contrasts, which I want to explore a little here. The first contrast is between models of phenomena and models of mechanisms, which doesn't quite, I think, map on to Prof. Fienberg's contrast between descriptive and generative models. The other contrast is between small networks which we know in rich contextual depth, and big networks where our knowledge is shallow and impoverished. Before elaborating on our divisions, however, I would like to say a few words about what we all seem to have in common. As the token representative of statistical physics on the panel, I will be deliberately provocative and say that what unites us is a devotion to the ideals of statistical mechanics.

Of course, of the participants at the workshop, only Dr. Clauset and myself were even arguably statistical physicists — and really he's a computer scientist and I'm a statistician. But the goal of statistical mechanics is to explain large-scale, macroscopic phenomena as the aggregated result of small-scale, microscopic behavior, as the result of interactions among individuals in contexts which themselves result from small-scale interactions among individuals. Global patterns should derive from local interactions. And this, I think, is something we would all be comfortable endorsing. Certainly when I heard Prof. Krackhardt explain that social networks matter because they show the contextual determinants of behavior, or when I saw Prof. Handcock check his ERGMs by seeing whether they could go from homophily and transitivity (local interactions) to the distribution of geodesic distances (a global pattern), my inner statistical mechanic felt right at home.[1] So, I'd claim that we're united by wanting to understand context and interaction, and how these lead to global patterns.

---

[1] To be sure, an inner economist, or an inner evolutionary biologist, would *also* have felt at home.

The first contrast, then, concerns *how* we do this. Once we have committed ourselves to *generating* the macroscopic patterns, we still need to decide whether we do this by modeling the *mechanisms* of action and interaction, and hope we get them right, or by modeling the consequences of interactions, and hope the details don't matter. The former leads to mechanistic models, the latter to what physicists call "phenomenological" ones. Take, for example, homophily. The model presented by Goldenberg and Zheng, for instance, is a mechanistic model, with a fairly detailed representation of the process by which people come to form social bonds, one consequence of which is homophily.[2] We also had phenomenological models of homophily, including both the ERGMs of Handcock and Morris, and the dynamic latent space model of Sarkar, Siddiqi and Gordon. Random walks in social space are obviously unrealistic, but may well be a good first approximation to reality; the ERGMs are simply silent about the dynamical processes by which networks form[3]. I *want* to have that a mechanistic understanding of the systems I study, so I find phenomenological models, as such, less than fully satisfying. But I recognize that there are very good reasons to use them, not the least of which is that they are much easier to get right. If Handcock and Morris want to measure the strength of homophily relative to transitivity, their problem is comparatively straightforward: estimate some parameters — with sufficient statistics, no less. If Goldenberg and Zheng want to make the same measurement for their model, the inferential problems are much more complicated, *because* their model includes mechanisms and not just phenomena.

The contrast between mechanistic and phenomenological models, then, seems to run through almost all the contributions here. But there is no reason we cannot have both sorts of models, or why we should think they contradict each other. In fact, I think this contrast is potentially productive of new research, since there should be ways of systematically deriving phenomenological models from mechanistic ones, and conversely of using well-estimated phenomenological models to constrain guesses about mechanisms.

I turn now to the second contrast, which is not between models of networks, but the networks themselves, or at least our representations of them. In small networks, like the karate club, or even the Colorado Springs sex-and-drugs network, we have, if not necessarily "thick descriptions" in the ethnographic sense, at any rate *deep* ones. We know a reasonable amount about each of the nodes, and sometimes (as in the karate club) can tell a story about each of the edges. We have, in other words, a lot of context, which is what we want. But, precisely because there is some much detail, it can be difficult, at a qualitative level, to distinguish an analytical narrative from a mere just-so story. If we then turn to quantitative models (which, as mathematical scientists, we're inclined to do

---

[2] In fact, they have what people in complex systems would call an "agent-based model". So far as I know, they are the only people to combine such a model with proper inference.

[3] The interesting paper by Hanneke and Xing adds dynamical detail to ERGMs, but makes no mechanistic commitments.

anyway), the small size of the network severely limits our ability to discriminate among models; the maximum attainable power is low.

Of course, we are no longer limited to small networks like the karate club; some of the graphs we saw presented at the workshop, like the PGP keyring network, had several million nodes, making them about five orders of magnitude larger than the karate club. This is exciting in its own right, because hitherto we have had almost no information about the *fine-grained* social organization of *large* populations. And certainly we no longer have much difficulty statistically distinguishing the predictions of different models! Dealing with this volume of strongly-dependent data does raise interesting technical problems; for instance, it's not obvious that models developed on small- and medium- sized networks can scale up, either computationally or descriptively, to such large networks. But beyond those technical problems, there is what seems like an intrinsic difficulty, which is that our knowledge of these large networks is shallow. It is simply not possible to have richly detailed information on all of the nodes, never mind all of the edges. When we look at *any* network with a few hundred thousand nodes, we are always going to be ignoring a huge amount of context about the nodes and their interactions. This isn't just a problem for social networks, but would also apply to, say, gene-regulatory networks.

So, opening up the divide between small networks and large, we find it contains a dilemma. Either we can possess the rich contextual detail we are interested in, or we can have enough data to severely test our models. Perhaps some clever methodology can cut a path through this dilemma; I myself don't see how.

Mark S. Handcock
Department of Statistics
University of Washington
Seattle, WA 98195-4322 USA
handcock@stat.washington.edu

The development of exponential family random graph models (ERGM) for networks has been limited by three interrelated factors: the complexity of realistic models, dearth of informative simulation studies, and a poor understanding of the properties of inferential methods.

The ERGM framework has connections to a broad array of literatures in many fields, and I emphasize its links to spatial statistics, statistical exponential families, log-linear models, and statistical physics.

Historically, exploration of the properties of these models has been limited by three factors. First, the complexity of realistic models has limited the insight that can be obtained using analytical methods. Second, statistical methods for stochastic simulation from general random graph models have only recently been become available [17,18,19]. Because of this, the properties of general models have not been explored in depth though simulation studies. Third, the properties of statistical methods for estimating model parameters based on observed networks have been poorly understood. The models and parameter values relevant to real networks is therefore largely unknown. Significant progress is now being made in each of these areas. However, despite their elegance and pedigree, the ERGM framework have yet to prove their value in addressing real scientific questions of interest. They have the tendency to produce degenerate behavior as a result of their maximum entropy properties [20]. This hinders simple model specification. The papers presented at the workshop illustrated many alternative approaches that may prove more fruitful.

The discussion of "generative" verses "descriptive" models was dialectic in nature. The exponential random graph models can be clearly be interpreted as descriptive. However, if we take the term generative to mean the ability to simulate network structures with given structural properties, they are also generative. If by generative is meant dynamic changing edges and structures then the paper of Steve Hanneke and Eric Xing illustrated how this can be achieved within the ERGM framework. If a probabilistic mechanism for adding additional nodes temporally is an regarded as an essential characteristic of a generative model then the published work on ERGM models does not meet this criterion. Note, however, that this is well within the capabilities of exponential family models. The model of Goldenberg and Zheng has a more directly generative mechanism and may be preferred for this reason.

The latent space framework invented by [3], and expanded by others at the workshop was originally descriptive in nature. However, variants of it can have a generative flavor (e.g., hierarchically adding a Gaussian mixture model for the positions).

As noted in the discussion, there are many challenges facing statistical network modeling. I believe the more traditional ones: inference from sampled data rather than a census, the development of statistical testing procedures, and their associated computational issues, will be overcome. The fundamental challenge is adapting the choice of models to the scientific objectives. Network phenomena are complex and the models must choose the specific features to be represented well while being ambivalent about the others.

Let me end by noting the success of this workshop in bringing together statistical network modeling researchers from distinct disciplines and scientific frameworks. The disciplines have much to communicate to each other especially where their scientific goals overlap. In the few cases were such researchers are brought together to speak, there has been little cross-disciplinary listening going on. This workshop was able to overcome that barrier so that researchers with backgrounds in SNA, physics, computer science or statistics were listened to. This success owes much to the principal organizers.

References

1. Frank, O., Strauss, D.: Markov graphs. Journal of the American Statistical Association **81**(395) (1986) 832–842
2. Strauss, D., Ikeda, M.: Pseudolikelihood estimation for social networks. Journal of the American Statistical Association **85**(409) (1990)
3. Hoff, P., Raftery, A., Handcock, M.: Latent space approaches to social network analysis. Journal of the American Statistical Association **97**(460) (2002) 1090–1098

4. Krebs, V.E.: Mapping networks of terrorist cells. Connections **24**(3) (2002) 43–52
5. Jordan, M.: Graphical models. Statistical Science **19**(1) (2004) 140–155
6. Handcock, M.S., Raftery, A.E., Tantrum, J.M.: Model-based clustering for social networks (with discussion). Journal of the Royal Statistical Society, Series A **170** (2007) in press
7. Stumpf, M.P.H., Wiuf, C., May, R.M.: Subnets of scale-free networks are not scale-free: Sampling properties of networks. Proceedings of the National Academy of Sciences **102**(12) (2005) 4221–4224
8. Rosen-Zvi, M., Griffiths, T., Steyvers, M., Smyth, P.: The author-topic model for authors and documents. In: Proceedings of the 20th Conference on Uncertainty in Artificial Intelligence. (2004)
9. McCallum, A., Corrada-Emmanuel, A., Wang, X.: A probabilistic model for topic and role discovery in social networks and message text. In: International Conference on Intelligence Analysis. (2005)
10. McCallum, A., Corrada-Emanuel, A., Wang, X.: Topic and role discovery in social networks. In: Proceedings of IJCAI 2005. (2005)
11. Wang, X., Mohanty, N., McCallum, A.: The author-topic model for authors and documents. In: Proceedings of the 20th Conference on Uncertainty in Artificial Intelligence. (2004)
12. Kemp, C., Tenenbaum, J.B., Griffiths, T.L., Yamada, T., Ueda, N.: Learning systems of concepts with an infinite relational model. In: AAAI. (2006)
13. Newman, D., Chemudugunta, C., Smyth, P., Steyvers, M.: Statistical entity-topic models. In: Proceedings of the 12th ACM SIGKDD Conference (KDD-06). (2006)
14. Airoldi, E., Blei, D., Xing, E., Fienberg, S.: Stochastic block models of mixed membership. Journal of Bayesian Analysis **(submitted)** (2006)
15. Zhou, D., Manavoglu, E., Li, J., Giles, C., Zha, H.: Probabilistic models for discovering e-communities. In: 15th International World Wide Web Conference (WWW2006). (2006)
16. McCallum, A., Pal, C., Wang, G.D.X.: Multi-conditional learning: Generative/discriminative training for clustering and classification. In: AAAI. (2006)
17. Snijders, T.A.B.: Markov chain monte carlo estimation of exponential random graph models. Journal of Social Structure **3**(2) (2002)
18. Handcock, M.S.: Degeneracy and inference for social network models. In: Paper presented at the Sunbelt XXII International Social Network Conference in New Orleans, LA. (2002)
19. Hunter, D.R., Handcock, M.S.: Inference in curved exponential family models for networks. Journal of Computational and Graphical Statistics **15** (2006) 1–19
20. Handcock, M.S.: Assessing degeneracy in statistical models of social networks. Working paper, Center for Statistics and the Social Sciences, University of Washington (2003)

Appendix

Statistical Network Analysis: Models, Issues, and New Directions
A Workshop at the 23rd International Conference on Machine
Learning (ICML 2006)
Thursday, June 29, 2006, Pittsburgh PA, USA

Schedule: (ALL SESSIONS IN RANGOS 3, UC 2nd FLOOR)

9:00-10:30 – Morning Session I

- Invited talk: *Heider vs Simmel: Comparing Generative Models of Network Formation*, David Krackhardt (Carnegie Mellon University)
- *A latent Space Model for rank data* , Isobel C. Gomley and Thomas B. Murphy (Trinity College Dublin)
- *Exploratory study of a new model for evolving networks*, Anna Goldenberg and Alice Zheng (Carnegie Mellon University)

11:00-12:30 – Morning Session II

- Invited talk: *Latent variable models of social networks and text*, Andrew McCallum (University of Massachusetts, Amherst)
- *Approximate kalman filters for embedding author-word co-occurrence data over time*, Purnamrita Sarkar, Sajid M. Siddiqi and Geoffrey J. Gordon (Carnegie Mellon University)
- *Analysis of a dynamic social network built from PGP keyrings*, Robert Warren, Dana Wilkinson (University of Waterloo) and Mike Warnecke (PSW Applied Research Inc.)

12:30-14:00 – Lunch and Poster Session – Joint with SRL and SOS Workshops

- *Stochastic block models of mixed membership: General formulation and "nested" variational inference*, Edoardo M. Airoldi (Carnegie Mellon University), David M. Blei (Princeton University), Stephen E. Fienberg and Eric P. Xing (Carnegie Mellon University)
- *Exploratory study of a new model for evolving networks*, Anna Goldenberg and Alice Zheng (Carnegie Mellon University)
- *A latent Space Model for rank data*, Isobel C. Gomley and Thomas B. Murphy (Trinity College Dublin)
- *Information marginalization on subgraphs*, Jiayuan Huang, (University of Waterloo), Tingshao Zhu, Russel Greiner, Dale Schuurmans (University of Alberta) and Dengyong Zhou (NEC Laboratories America)
- *Predicting protein-protein interactions using relational features*, Louis Licamele and Lise Getoor (University of Maryland, College Park)

E.M. Airoldi et al. (Eds.): ICML 2006 Ws, LNCS 4503, pp. 195–196, 2007.
© Springer-Verlag Berlin Heidelberg 2007

- *Age and geographic inferences of the LiveJournal social network*, Ian MacKinnon and Robert Warren (University of Waterloo)
- *A brief survey of machine learning methods for classification in networked data and an application to suspicion scoring*, Sofus A. Macskassy (Fetch Technologies Inc.) and Foster Provost (New York University)
- *Inferring formal titles in organizational email archives*, Galileo M.S. Namata Jr, Lise Getoor (University of Maryland, College Park) and Christopher P. Diehl (John Hopkins Applied Physics Laboratory)
- *Approximate kalman filters for embedding author-word co-occurrence data over time*, Purnamrita Sarkar, Sajid M. Siddiqi and Geoffrey J. Gordon (Carnegie Mellon University)
- *Discovering functional communities in dynamical networks*, Cosma R. Shalizi (Carnegie Mellon University) and Marcelo F. Camperi (University of San Francisco, San Francisco)
- *Learning approximate MRFs from large transaction data*, Chao Wang and Srinivasan Parthasarathy (Ohio State University)
- *Entity relationship labeling in affiliation networks*, Bin Zhao, Prithviraj Sen and Lise Getoor (University of Maryland, College Park)

14:00-15:30 – Afternoon Session I

- Invited talk: *A review of statistical models for networks*, Stanley Wasserman (Indiana University)
- *Discrete temporal models of social networks*, Steve Hanneke and Eric Xing (Carnegie Mellon University)
- *A simple model for complex networks with arbitrary degree distribution and clustering*, Mark S. Handcock and Martina Morris (University of Washington, Seattle)

16:00-16:30 – Afternoon Session II

- *Strutural inference of hierarchies in networks*, Aaron Clauset, Cristopher Moore (University of New Mexico, Albuquerque) and Mark Newman (University of Michigan, Ann Arbor)

16:30-18:00 – Closing Session

- Invited panel discussion
    Stephen Fienberg, Chair (Carnegie Mellon University)
    David Blei (Princeton University),
    David Krackhardt (Carnegie Mellon University),
    Andrew McCallum (University of Massachusetts, Amherst),
    Cosma Shalizi (Carnegie Mellon University),
    Stanley Wasserman (Indiana University)

- Closing remarks

Author Index

Lecture Notes in Computer Science

For information about Vols. 1–4534

please contact your bookseller or Springer